THE PHYSICS OF CRIC

Nottingham University Press
Manor Farm, Main Street, Thrumpton
Nottingham, NG11 0AX, United Kingdom
www.nup.com

NOTTINGHAM

First published 2011
© Mark Kidger

British Library Cataloguing in Publication Data

The Physics of Cricket
M Kidger

ISBN 978-1-904761-92-1

Cover photo adapted from www.morguefile.com

Typeset by Nottingham University Press, Nottingham
Printed and bound by Berforts Group, Hertfordshire, England

The Physics of Cricket

Mark Kidger

Nottingham
University Press

TABLE OF CONTENTS

INTRODUCTION

How did this book come about? Ultimately it was very simple, but even simple things can sometimes hide a long story.

If you are reading this book it is a strong bet that, like me, you were born in or live in one of a small number of former British colonies that took up the game of cricket in the late 19th or early 20th century. To people in other countries, cricket is regarded as a peculiar British pastime although, in fact, the vast majority of players and cricket fans live in the Indian sub-continent, or are of Asian descent. Surprisingly though, because of this huge Asian interest, the Cricket World Cup is the most watched sporting event in the world after the football World Cup and the Olympic Games.

Cricket is, in fact, played in around 100 countries around the world, many of them completely unexpected. As I write, as well as games between New Zealand and India and between South Africa and Australia, the final qualifying tournament for the 2011 Cricket World Cup in Bangladesh, India and Sri Lanka is under way. Today's matches, which are all taking place in South Africa, have seen Kenya beat Afghanistan, Bermuda beat Denmark and Canada beat Namibia. Previous rounds have seen teams such as Argentina, Papua New Guinea, Fiji, Israel, the United States and Italy eliminated. Dutch and Danish cricketers play professionally in England. Cricket has even featured once in the Olympic Games when, in the 1900 Olympics in Paris, an English team called the Devon and Somerset Wanderers beat their only rivals, a French team (of mainly British expatriates) for the Gold Medal, after the Dutch and Belgian teams withdrew![1]

The poster announcing what turned out to be the 1900 Paris Olympics cricket Gold Medal match.

As we will see in more detail in Chapter 1, cricket had been played internationally long before the 1900 Olympics. In fact, it had even been played long before what is now recognised as the first formal Test match, as games between the most powerful teams are denominated. Held in Melbourne between Australia and England, the first Test match started on March 15th 1877 and was won, by a quite narrow margin, by the Australians. This established a dominance over the English that has been maintained, with certain breaks, ever since. The very first international sporting fixture in the world and the longest series of matches in any sport is the regular cricket match between the United States and Canada, which started in 1844 with a Canadian victory. It seems hard to believe now but, up to the late 19th Century, cricket was played to a high standard in the United States and matches there drew big crowds.

I have been passionate about cricket in all its forms, including following the scores from the lesser outposts of the cricketing world since childhood and can remember, when I was seven years old, regaling my father on his arrival home with a full description of the action from one day's play at Headingley, knowing that he was interested. My primary school played organised cricket and had regular matches through the summer and when I was not playing for the team and revealing my failings[2], I was travelling to matches as scorer and practicing after school, or playing in the garden at home, either alone or with my brothers. Hard balls were too dangerous in a small space with windows so, often we used roughly shaped balls made from newspaper and copious quantities of sticky tape. Later, as my brothers grew up and had their own pursuits, I would pass many hours practicing bowling on my own (usually using a plastic, airflow ball; one of the tricks with these was to put it in the freezer before playing – until it warmed, it would be rock-hard and have good bounce, thus imitating a new ball in cricket).

As a cricketer I suffered from one serious physical handicap that reduced my value to any team: the counterpoint to my enthusiasm was an innate lack of talent! Despite this, I played for multiple teams over the years, from my primary school, through my House team at Bristol Grammar school, the school 3rd XI – a team reserved for those who were similarly enthusiastic, but shared my lack of ability, albeit usually to a lesser degree than me – and then for the staff team at the bakery where I worked during the summer break while at university, indoor cricket at university and then for the staff team at the school where I was teaching at East Malling in Kent and one of the local village sides (Addington, http://en.wikipedia.org/wiki/Addington,_Kent) . In my last season before leaving to work in Spain I even started to show some signs of modest ability, taking around 50 wickets in the season and regularly being the best bowler that school staff team and the guy who they turned to for wickets and to win them matches. When I was not playing I was keeping the score and producing match reports. Of course, I would love to presume about the afternoon when I took my best ever bowling figures of 4-12 from 5 overs but, it was against the school XI and our captain declared, with sadistic relish, that if I expected to be allowed to bat too the match would have to be declared "not first class" and so not count towards my season's figures (the brute!!! I didn't get many chances to bat and even fewer to shine with the bat, so this dire threat cut deep!) Moments of glory with the bat in my hand were rare; my highest ever score was 26, which I managed twice – not the stuff to terrorise the Australian bowling attack – but, at our level, even I could have the occasional moment of glory[3]. These were happy games. The staff matches were played usually midweek evenings after work and, frequently on Saturday and Sunday I would play for Addington all afternoon. This was cricket at its best: social, frequently laughing at our own incompetence, talking, enjoying fresh air and countryside and, unlike after a rugby match, people would drink just enough to be sociable and no one would comment on me supping with lemonade.

The village matches would feature the traditional cricket tea: everyone around one or two big tables, knocking back cake, sandwiches and tea. For the genuine village cricketer the tea in the middle of the match is almost more important than the match itself and the hosts would always do their best to excel with their hospitality, even when we were only going to enjoy it for twenty minutes. In recognition of this tradition, at one point the BBC even organised a competition to see who prepared the best cricket tea from the many ladies (and a few men) who would give up their weekend afternoons to prepare teas for hungry and appreciative players.

This last is one of the keys to cricket. It is a social game. People go along to enjoy themselves. Matches are almost invariably played in a gentlemanly manner, with dissent rare and strongly frowned upon. And you can go to a big match in a full stadium with intense rivalry between the competing sides and expect not to be bothered by rival fans and to be able to get home alive afterwards. When at school I would think nothing of going on a Sunday afternoon to watch my team – Gloucestershire – play in a packed ground without a policeman in sight and, after walking a couple of miles from the nearest bus stop to get to the ground, walk 6 or 7 miles home afterwards because buses were so infrequent on a Sunday evening. And this at a time when no sane person would want to be walking around outside the nearby football stadium on a match day.

As the possibility of playing organised cricket disappeared, first the radio (blesséd be the name of the BBC World Service) and then the Internet, supplemented by books and magazines, fed my voracious appetite for the game. Hardly a day of the year passes without some game of cricket being played somewhere around the world. Now the Internet allows one to follow the score, as it changes, whether the game is being played in Auckland or Edinburgh, Cape Town or Barbados, Bristol or Bangladesh. In the 1950s the radio allowed fans to follow their heroes from thousands of miles away, imagining their deeds described in the silvery words of the commentator. Now, increasingly often, the Internet allows one to listen to commentary from an ever-widening number of matches. Over the years I have become totally addicted to following the fortunes of Middlesex, described first by BBC Radio London's legendary Norman de la Mesquita and, later, by the equally legendary Ned Hall. In recent seasons Kevin Hand has taken up the baton and does it with a style and sense of fun that is so typical of radio cricket commentary. Kevin's unique mixture of commentary and interaction with listeners around the globe livens up even the dullest day[4]. Over the years Kevin and I have exchanged many emails. His open forum on the radio has given me a chance to write about the game that I love and interact with other fans.

Cricket, in its purest form, is a very relaxing game. In the traditional form it is played at a gentle pace and takes several days to complete a game, often without even a positive result. Like in chess, an unfinished game is called a draw, even if one player or side has a huge advantage over the other when the allotted time ends. As the legendary John Arlot, the BBC poetry producer who became possibly the world's greatest ever cricket commentator, said "it's a contemplative game". There is time to think, but a moment's loss of concentration can lead to disaster. Deep tactics and strategy are involved as well as large doses of psychology. Perhaps this contemplative, unhurried aspect is the reason why cricket is deeply linked to poetry, literature and art. Cricket has led to entire anthologies of poetry in an intellectual outpouring that would be unthinkable in football, as well as being the civilised, thinking academic's game.

The following is one of the best known of all verses in English poetry, encompassing the British spirit in the age of Empire:

VITAI LAMPADA
("They Pass On The Torch of Life")

There's a breathless hush in the Close to-night --
Ten to make and the match to win --
A bumping pitch and a blinding light,
An hour to play and the last man in.
And it's not for the sake of a ribboned coat,
Or the selfish hope of a season's fame,
But his Captain's hand on his shoulder smote --
'Play up! play up! and play the game!'

In the poem, written in 1897, Sir Henry Newbolt wrote of how a future soldier learned to be stoic by playing cricket in the famous close at Clifton College in Bristol. The two following verses describe the horrors of war and how the values of cricket would help the boy to bear them. Although the modern generation would recoil in horror at the jingoistic sentiments, our fathers or grandparents, who had fought and, all too often, laid down their lives in the Second World War, would have understood them perfectly. The team spirit of cricket where one hopes in a selfish way to shine individually, while at the same time the captain has to ensure that the whole is greater than the sum of the parts, has always been a favourite for encouraging unity of purpose. It is not for nothing that the young Douglas Bader, later to become the RAF's great legless fighter ace, was advised before his interview, when seeking to become an RAF cadet, that he should stress his particular love of sport and, especially, of team games such as rugby and cricket.

In the same way, as it has inspired poets, cricket and cricket matches often attract artists. It is not unusual to find someone with the easel and canvas painting the scene at a match. The fact that games last several days makes them attractive targets for artists who need time to take in the colour, the detail and the background. Probably cricket has been captured on canvas more than any other sporting theme over the years.

So, cricket and the arts are often associated, but cricket and science? Despite the way that it has attracted intellectuals, cricket has not been an overly scientific game: you are far more likely to find a poet than a scientist in a lab coat wandering around the ground. But cricket is actually increasingly dependent on science. The bat, protective equipment, even the ball are increasingly high technology. And, in the last few years, technological aids have increasingly both helped and haunted the men in white in the middle who control the game, scrutinising and helping to adjudicate their decisions. Cricket is also, *par excellence*, a game for the mathematician and statistician.

For years I had been thinking about writing a book about cricket. What though would be my slant on it? I have never played at any serious level. I have not covered games in the press, so how could I write anything new or authoritative that might be worth publishing? Yet still the idea appealed to me, an itch that could not be scratched. Finally, like a bolt out of the blue, came an e-mail from an old friend and erstwhile editor of my astronomy books, Trevor Lipscombe, trumpeting triumphantly his first published book "The Physics of Rugby" and wondering out loud if I might be interested in writing "The Physics of Cricket". The more I thought about it, the more the idea appealed and, thanks to Trevor and to the friendly team at Nottingham University Press, here it is.

In the Internet forum that I participate in, discussing cricket, people often make a point and say "it isn't rocket science!" Mischievous fortune has ensured that I have, quite genuinely, become a rocket scientist (although my interest is in what goes inside the pointy bit at the top, rather than the loud, firey bit underneath). Behind cricket there is a huge amount of science that player, commentator and spectator must understand, if only empirically. Whether it is the Pakistani bowler, Shoaib Ahktar sending down his own rockets at 161km/h (in old units, the legendary 100 mile per hour barrier, which seemed as unbreakable as the Sound Barrier did to the aviators of the late 1940s), or Ian Botham seemingly launching the ball into orbit with his bat, rocket science or otherwise, there is a lot of science involved: optics; mechanics, fluid dynamics; materials science; statistics; infrared technology;

acoustics; anatomy; etc that the players often use unconsciously, but use it they do. Whatever your interest, sit back and enjoy the ride as we explore a little of how our heroes do what they do so well and how the laws of physics and mathematics that allow them to do it... or sometimes not.

Trevor has done a marvellous job with "The Physics of Rugby", due in no small part to his love of that game his and experience of playing it at university level. I make no apologies for imitating his style and trying to make this a worthy companion volume to his.

Endnotes

[1] There have been efforts in recent years to get cricket included in the Olympic Games once again, having featured in the Asian Games and the Commonwealth Games in recent editions. As I revise this text, cricket has just been cleared for a potential application for inclusion in the Games after 2016, although attempts to get it included in the 2012 London Games, in its T20 format, as a demonstration sport, have failed. Lords cricket ground will feature in the Olympics, but as an archery venue.

[2] After my third consecutive duck I asked the games teacher nervously if I would be dropped. He looked at me darkly and growled "from a very great height".

[3] As you ask, I will recount just two; one intentional, the other not! Just once I won a match with a doughty batting display at number ten. Our side was chasing a small total and making a bit of a horlicks of it. I came in to bat with about a dozen still needed and only the world's greatest batting rabbit to come (don't worry Graham, I won't reveal your name). Very rapidly the batsman at the other end played a wild shot and left us with nine to get and no expectation of having any chance of getting them. There was also the not inconsiderable problem that our number eleven hadn't scored a run for several matches and was facing the bowling. There was only one thing for it and that was to get to his end somehow and face the bowling myself! A suicidal single and a misfield did it and, slowly the runs were whittled down. Towards the end of the next over the spinner floated one up and I went down on one knee and swept the ball elegantly for the winning boundary. I am not sure if it was the bowler, my team-mates, or me who was most surprised – I didn't know that I could play the sweep shot!!! Disappointingly, even after this fine performance, the England selectors did not call me.

The other moment was my only six in competitive cricket. In a game against a local club – West Malling, although playing under a different and somewhat unprintable name (for the true cricket buff West Malling is a historical name, as it was home to one of the cricket grounds of the 18th and 19th Century: Fuller Pilch's Ground). The game was played in the most awful conditions, It was dark, it was wet, it was muddy, it was drizzling for much of the match and it was utterly miserable to the extent that a couple of our best players did not turn up, as they were expecting the game to be called-off. For some unfathomable reason I was asked to bat at three and was not doing too badly. Suddenly, a head-high ball came at me out of the gloom. My first thought was to duck out of the way, but then I realised that if I did it would probably follow me and hit me. So, I decided to stand my ground – the fact that I had time to think all this gives you an idea of how slow the ball was – and swing like hell. Miraculously, not only did my hook shot connect, but it flew over the Fine Leg boundary for a quite magnificent six. I would love to say that it was deliberate.

4   This style of commentary was pioneered by the BBC's "Test Match Special" (TMS) programme, broadcasting England's matches to the point that listeners even wrote in to say that the commentary was even more entertaining when there was no play in the match to interrupt the chat and by-play. Unlike the TMS team though, where as many as six commentators will take turns through the day, with always at least two on the microphone at any time, the amazing Kevin Hand broadcasts for six hours a day alone, showing a professionalism and devotion to his job that is peerless. Throughout the day, listeners from around the world will email him and comment, analyse, crack jokes and joust verbally (and yes, sometimes they will also tease him mercilessly, usually producing stifled giggles as he tries to continue). His pride in announcing that a regular listener from Tokyo (or Thailand, or Korea, or Washington State) has just written in for the first time in a new season is a wonder to behold. These commentaries show that the first thing in cricket is that it is *fun*, however much one wants one's own side to win. You can find a link to Kevin's commentary on the BBC Sport Website throughout the summer (http://news.bbc.co.uk/sport2/hi/cricket/default.stm).

Chapter 1

A short history of cricket

THE IMAGE OF IPOCRISIE

O lodre of Ipocrites,

Nowe shut vpp your wickettes,

And clappe to your clickettes!

A! Farewell, kings of crekettes!

Poem (probably incorrectly) attributed to John Skelton

In 2009, a lecturer at the Australian National University called Paul Campbell, drew attention to the poem above. Supposedly written by English poet, John Skelton (c 1460-1529) in 1533 – there is an obvious inconsistency here in the dates leading to a real doubt that the author really was Skelton – who became Poet Laureate. Skelton was educated in Leuven in Belgium, as well as at Oxford and Cambridge and so had a close relationship with Flanders. The poem reputedly describes the Flemish weavers who settled in the south and east of England in the 15th Century. There are two striking words in the verse that appear to be cricketing terms "wickettes" (wickets) and "crekettes" (cricketers?) It seems that the weavers are referred to in the verse as the kings of cricket; a rather unexpected turn of events! It has been suggested that this verse is proof that cricket was not an English invention after all, but instead brought to England from modern-day Belgium[1]. It must be mentioned though that other sources are sceptical of these poetic references, pointing out that the verse appears to be more an attack on the Pope than a reference to cricketing Flemish weavers and note the inconsistency in dates between that attributed to the poem and the poet's own date of death[2].

It has long been speculated that cricket as a game possibly began with shepherds using their crooks as bats, defending themselves against stones or wooden balls (although one suspects that this latter was a rare luxury in improvised games), rolled against the wicket fences that held their sheep. This would have been a leisure activity, carried out when things were relaxed and simply depended on materials that were always to hand. The playing surface was not flat – any piece of bare ground or short grass big enough for half a dozen or more men would serve – and the shape of the bat used obliged the improvised ball to be rolled along the ground rather than pitched, hence the term *bowling*. Given the thinness of the part of the crook that would be in contact with the ground, any delivery that bounced more than a few centimetres off the surface would be almost unhittable; one can imagine that any kind of irregularity in the ground or the improvised ball would lead to a lot of cursing and bruises as unprotected ankles and shins took the impact of blows. Presumably there was a gentlemens' agreement not to maim opponents deliberately while playing!

Another widely supported popular theory suggests that cricket originated out of a childrens' game. What is clear is that the first clear historical reference to cricket is of it as a school field game and dates from 1597. In a court case held in that year, it was testified that "creckett" had been played fifty years earlier on a disputed plot of land, by children of the Guildford Royal Grammar School. This statement was used to back the school's claim of ownership of the land. Earlier references to cricket are very vague. Apart from the poem above, which suggests that cricket was being played early in the 16th century, there is a reference in 1300, in the accounts of King Edward I of England, to a game called "Creag", played by Prince Edward for which payment was made. There is though no information whatsoever to suggest what kind game "Creag" might have been, so the only association with cricket is purely and exclusively a loose phonetic one that is regarded as having limited credibility[3].

There are many mentions to cricket by the early 17th Century, including the first records of organised matches by teams from different villages. Cricket seems to have been a somewhat hazardous occupation at that time, because the references to the game mostly involve litigation, or report serious injuries due to cricketing accidents. There are references to several court cases involving playing cricket on a Sunday, including one in which two men were given a stiff fine and ordered to do penance for having missed church to play cricket. Worse, there are various references to people maimed or killed by cricket bats; it may be this that led to an important change in the rules of the game when they were set down formally for the first time in 1744. Up to 1744 it was accepted practice that if the batsman hit the ball into the air he could hit it again to avoid being caught. This led to several instances where a batsman swung wildly to impede a catch and hit the lunging fielder a fatal blow. In the 1744 laws, presumably to avoid further accidents of this kind, the practice of hitting the ball twice was outlawed and became a mode of dismissal instead.

In England, the 17th Century was marked by the English Civil War, which ended in 1646 with the defeat of the Royalist forces at the hands of Oliver Cromwell. About this time cricket took off as a popular pastime. Part of the reason was that it was a highly social game, well suited to large country estates, where staff and their masters could join in a common cause, thus fomenting a team spirit. Another reason was that the games made excellent opportunities for betting. Undoubtedly, it was the large amount of gambling that took place that served as a powerful motor for competitive cricket. As the ruling regime became more relaxed and less puritan, gambling on cricket became an increasingly important industry and served as an incentive to raise the level of play; rich patrons wanted to increase their winnings by forming their own, select teams of the best players. This led inevitably to cricket ceasing to have as its highest level matches between neighbouring villages and turning increasingly to sides drawn from whole counties or regions. It also led to the first generation of professional cricketers, employed by a patron to play exclusively for his side[4]. Even if an Act of Parliament at that time limited wagers to £100, this was still worth upwards of £10 000 at 2010 prices and thus was a quantity that was taken very seriously indeed. A winning team could be an important source of wealth for a patron.

As cricket became more important and involved ever-larger sums of money, it became necessary to have a consistent set of rules by which it could be played. Nothing is more guaranteed to cause controversy and conflict than a large bet placed on a game with vague rules that are open to widely differing interpretations, especially when one side clearly has the upper hand. The first attempt at drawing-up such laws was drawn up by the nobles and gentlemen who played at the Artillery Ground in London. Although not universally followed, the laws of 1744 set down the basis of the modern form of the game, such as the length of the pitch and the size of the stumps.

Various attempts were made to refine and reform the laws over the following decades, but the key date in cricketing history was May 30th 1788, when the recently formed Mary-le-Bone Cricket Club produced its first set of laws, standardising most aspects of the game in a single code that, progressively, became accepted as THE laws of the game.

The Marylebone Cricket Club, or MCC as it is better known, became the custodians of cricket, a role that it kept until 1993, when the International Cricket Council (ICC) took over[5]. Even so, touring English teams still competed under the MCC banner until 1997, when the last MCC tour, of New Zealand, ended. Whereas England sides had played as "England" in Tests and One Day Internationals and as "MCC" in other tour matches, it now simply became an "England XI" for all non-international tour matches[6].

The Marylebone Cricket Club took the best part of a century to become the focal point of the game. At the end of the 18th Century, as cricket became an increasing important pastime in England, the dominant force in cricket was a small village on the south coast. Hambledon could boast a side so strong that, in 1772, it could take on and beat in an odds match – i.e. 11 Hambledon players against 22 opponents – a full England side. In 1777, Hambledon played England again, this time at Sevenoaks Vine, winning by the overwhelming margin in those days of typically low scores[7], of an innings and 168 runs.

The Hambledon matches were played habitually for a purse of £500: about £50 000 at 2010 prices. Considering that Hambledon played numerous matches, the quantity of money involved in their matches was huge.

In his book "Cricket – A Way of Life[8]", Christopher Martin-Jenkins comments:

> *That this otherwise unexceptional village should have become prominent
> in cricket between about 1760 and 1785, and the undisputed centre of the
> game in the 1770s, is one of the more delightful accidents of English social*

history. It happened that there were a good many good cricketers in the area, a few keen patrons, and in Richard Nyren a man with passion for cricket and an administrative flair which enabled him to form elevens capable of taking on and beating the best team that 'All England' could muster.

Sadly, it was this very quaintness of Hambledon that was its eventual downfall. In the days before railways, when transport between towns was a major endeavour, a pitch on a windswept moor by an isolated village was not the place where the great and good could conveniently come to lay their wagers on matches. In a way, Hambledon's very success led to its being superseded as a venue. London, with its rapidly increasing population and villages closer to London such as Sevenoaks and West Malling, became the venue of choice for the increasing number of important matches played for major wagers.

The first match at Lords Cricket Ground, on its present site of St. Johns Wood, in north London, was played in 1814. In it, the MCC beat Hertfordshire by an innings. By now the MCC was already becoming an important and increasingly authoritative voice of cricket.

Until the mid-nineteenth century cricket was very much limited to the Home Counties and Hampshire – effectively the south and southeast of England. At this time though, the transport problem that had eventually stymied Hambledon's role was solved. The invention of the railway made travel around the county faster, more comfortable and more efficient. This led to what was effectively the first travelling circus of cricket, as William Clarke, a professional from Nottingham, formed an England side that he led around the country playing mainly so-called "odds" matches, usually against local teams of twenty-two players. Crowds that were frequently around ten thousand, meant that the games were a profitable enterprise. Ten thousand spectators paying two pence each at the gate meant total receipts of over £80, allowing the players to be paid generously and still leave a healthy profit for William Clarke.

At the same time, traditional matches such as the Gentlemen v The Players started and rapidly became important fixtures in the cricketing calendar. The scene was set for two hugely important developments. In the south, the MCC was laying the foundations for what would become county cricket. And, as tours by select teams demonstrated that they were popular and thus profitable, it became inevitable that someone would propose an international tour: the foundation for what would soon become Test cricket was laid.

In the 1850s, the main powerhouse of cricket outside England was not Australia, or South Africa, or the Indian sub-continent: it was North America. The lack of any credible rivals with sufficient proximity to England made the idea of England playing an international match unrealistic. This was not true though in North America. Cricket was being played to a good standard in both the United States and Canada. In 1844 the two met for the first international fixture of any kind. This match probably has as good a claim to be the first ever Test match as the later games between England and Australia but, historical accident has not given it either the status, or the recognition that the match and its successors deserved. Few people realise that the oldest international sporting fixture of any kind is not England v Scotland at football (the first such fixture was played in 1870), or Scotland v England at rugby (again, the first match dates from 1870), but instead it is the United States v Canada in cricket, a game that is still played almost every year by these two oldest of all sporting rivals.

The inaugural match was played in New York from September 25th-27th 1844. Initially intended to be New York v Toronto[9], it was billed instead as USA v Canada, with the American team being drawn from the big cities of the east coast. The match was extended to a third day after the second day was completely rained-off (thus initiating another great cricket tradition!) Canada, with 82 and 62, ran out comfortable winners by 23 runs as the United States could only reply with 64 and 58; first innings advantage, as so often, proving decisive in a low-scoring game. It seems hard to believe now that this match attracted a huge crowd: the attendance was variously estimated as being as high as twenty thousand over the two days of play.

Just as the railways opened up England for cricket tours, the advent of steam ships brought the colonies closer and made international tours inevitable. In 1859 the Montreal Club invited an England side to visit Canada and the United States. England sent an extremely strong side. Each of the games was an "odds" match; had any of them been eleven against eleven it is possible that it would later have been recognised as the very first Test match. However, the match against the 23 of the USA and Canada at Rochester on October 21st, 24th and 25th was miserably one-sided. The USA and Canada batted first and struggled to 39 off 35 four-ball overs. John Jackson and John Wisden bowled unchanged, with Wisden taking the quite incredible figures of 16-18 from 18 overs. Only one of the last ten USA and Canada batsmen troubled the scorers. England replied with 171, with Tom Hayward making 50 – the only batsman in the match to reach 20 – although the wonderfully named Julius Caesar "disappointed" by scoring just 11 batting down the order at nine. Jackson and Wisden again bowled unchanged in the second innings, this time knocking over the opposition, who batted two men short through injury, for 62, with Wisden this time taking 13-44, for a match analysis of 29 wickets for 62 runs.

During the second half of the 19th Century the Philadelphians played cricket to First Class standard. They made three tours to England between 1880 and 1897 and enjoyed some real success against county sides, showing what potential was missed as cricket died in North America.

Unfortunately, by the time of the Philadelphians final tour in 1897, which featured wins against Sussex and Warwickshire, cricket had already been supplanted as the leisure game of Americans at play. It would be pointless to speculate what might have happened had the American Civil War not broken out in 1861, although it is worth remembering that the Americans were playing at a level and in circumstances not dissimilar to Australia's at the birth of Test cricket a few years later, although it has been argued that even then the grassroots structure of the game was stronger in Australia. It is fair to say that had circumstances been slightly different and American cricket a little luckier, the first ever Test match could well have been a game between the USA and England, played in the 1860s or 1870s. However, unfortunately for the development of the sport, the Civil War severely disrupted commerce and communications, both internal and external. Cricket, at any serious level, requires considerable effort and investment: a cricket ground must be cared for carefully and both playing and maintaining a ground requires a lot of specialist equipment. Troops at play in the Civil War turned to the much simpler requirements of baseball, for which a stick, a ball and any patch of fairly level ground would suffice[10]. By the time that the Civil War had ended, the opportunity for cricket in the USA to become one of the leaders of the world had gone and baseball became the game of the masses, with cricket limited to the social elite. With the death of the cricket-playing and supporting social elite, cricket in the USA became very much a minority sport, even though the USA v Canada matches did continue on a more or less annual basis[11].

With the American Civil War also more or less ending the appeal of the North American continent as a touring destination, English cricketers looked elsewhere: the Australian colonies were in the right place at the right time. Cue the start of perhaps the greatest historical rivalry in international sport.

The first England tour of Australia happened though by accident, not design. A reading tour by Charles Dickens had been arranged, but fell through and a cricket tour was hastily put in place as a substitute. Cricket was firmly established in Victoria and New South Wales and no less than twelve matches were played in front of often big crowds. England's record was P12, W10, L2, but both defeats came in odds matches against 22s. In 1873/74 W.G. Grace's side played 15 odds matches, winning only 10 of them. There was no question that the colony of Australia was catching-up fast with their colonial masters.

By 1876 there had been three tours of Australia by England and one tour of England by an Australian aboriginal side. James Lillywhite led a side on English cricket's fourth foray to the Antipodes. Note though that, on this occasion, amongst other missing stars because Lillywhite picked only professionals, the Grace brothers were absent, despite WG topping the averages for the 1876 season and, to boot, having more than double the run aggregate of any other batsman, more than twice as many centuries and the season's highest score – a *mere* 344. In 1874 WG had also taken a small matter of 140 wickets at 12.70, with 191 at 12.94 in 1875 and would manage 179 at 12.81 in 1877! In fact, by not inviting any amateurs at all, Lillywhite had deprived himself of no less than 19 of the top 20 batsmen in the 1876 English averages.

A portrait of James Lillywhite's team to tour Australia and New Zealand in 1876/77 taken shortly before leaving England. Back: Harry Jupp, Tom Emmett, Alfred Hogben (a sponsor of the trip), Allan Hill, Tom Armitage. Front: Ted Pooley, James Southerton, James Lillywhite jnr, Alfred Shaw, George Ulyett, Andrew Greenwood. On ground: Harry Charlwood, John Selby.

Once again the side played almost exclusively odds matches[12] but, after a trip to New Zealand, a match against a combined Australian XI, all from New South Wales and Victoria, was hastily added to the schedule of a tired England team who had suffered greatly from the rough sea crossing from New Zealand, arriving back in Australia just 24 hours before the game started[13]. Although not billed as a Test match at the time, this first match on even terms between the two representative sides was retrospectively given Tests status. The match started on March 15th 1877 in front of a small crowd of just 1500 in a ground that even then could hold 30 000. In a game where no other Australian batsman passed 20, Charles Bannerman's monumental score of 165 before retiring hurt, set up an Australian win by 45 runs[14] when England, chasing 155 for victory, were shot out for 108, with English born Tom Kendall taking 7-55. Although Lillywhite's side gained revenge with a comfortable win in a second match, the honour of victory in the first ever match on even terms went to the home side.

Tours of Australia at the time were, as can be seen, always arranged as private ventures, led by a particular player and selected by invitation, not as formal representative sides selected by a central body. There was no pretence that any of the sides represented the best players in England and not all of them were willing to make the gruelling journey that could last six months even if invited to do so. This led to widely differing interpretations as to what constituted a Test match between the Australians and the English, with the former recognising more matches as having been worthy of Test status. However when, finally, agreement was arrived at on the formal list, it was Australia's view which prevailed eventually in the early 20th Century[15].

It was not until the 1903/04 series, the 22nd between the two sides and no less than the 12th tour of Australia in the modern list of Tests, that finally selection of the touring party came under the auspices of the MCC. The effects of this self-defeating selection can be seen in the relative fraction of wins for one and the other side in Tests in England and Australia. Despite England winning three consecutive series and four out of five in Australia between 1884 and 1895, Australia's win-loss record at home was 20-15 (almost all Tests were played to a finish in Australia, so there were just 2 draws in this sequence). In contrast, in England, with fewer limits on selection, England had won 13 of the 19 completed matches, with 10 draws. However, it was one of the six defeats in England that was the spur for England to take the challenge of Australia far more seriously than before.

Before 1882, just one match had been played between England and Australia in England, as against seven matches over three series in Australia. Thanks in no small measure to an imperious 152 from W.G. Grace, England had won the 1880 match at The Oval comfortably. When Australia arrived in 1882 to play another Test at The Oval no one took their challenge too seriously and with Australia rolled over quickly for 63 on the first day, it seemed that another easy win was in the offing. Withstanding the efforts of Fred "The Demon" Spofforth, England scraped past 100 and established a seemingly decisive first innings lead of 38. Despite Hugh Massie's heroic second innings 55 – his only half century in his nine Tests – England were set a mere 85 to win and, at 51-2 with W.G. Grace well set and Geoff Ulyett giving him solid support, a win seemed just a formality. Little did the crowd suspect just how premature their celebration of the impending victory would turn out to be.

In the space of a few deliveries Spofforth dismissed Ulyett to a catch behind and Harry Boyle induced a catch from the seemingly impregnable W.G. but, as Lucas and Lyttelton took the score to 66-4 with just 19 needed to win, Grace's dismissal seemed likely to be no more than a minor consolation for the Australians. Suddenly though Spofforth started to bowl like the demon that he was nicknamed. He clean bowled Lyttelton, 66-5. He caught and bowled Steel, 70-6. He bowled Read first ball, 70-7. And finally, bowled Lucas, the last remaining specialist batsman, who had stuck out there in the middle for more than an hour for just 5 runs. The score was now 75-8. England needed just ten runs to win, but Australia had their tails up and required just two wickets. Barnes lasted just 5 balls and Peate, 3 as Harry Boyle dismissed both to round off the victory, with England all out for 77, 8 runs short.

England was stunned. The country could not believe that their heroes had been beaten by the Australians. The following day The Times famously ran its obituary mourning the death of English cricket, stating that it would be cremated and its ashes would be taken to Australia, from which phrase the famous trophy took its name[16].

A stunned nation was shaken out of its complacency for a while and of the next twelve series between the two sides, England won ten, the only blemishes being a 2-2 draw in Australia in 1882/83 and a 2-1 defeat in Australia in 1891/92[17].

In 1888 a second nation started to play Tests when a much-depleted England side visited South Africa. It was almost 20 years before the South Africans were regarded as strong enough to send a team to England and, however weak the England side that was picked, the first series were still almost ludicrously one-sided: four of the first five matches ended in innings victories for England, who won every single Test of the first four series. Australia though managed a dominance over South Africa that was insulting in the extreme. Between 1902 and 1950 there were seven series between the two sides, with Australia winning 22 matches, drawing 6 and losing just 1 match[18].

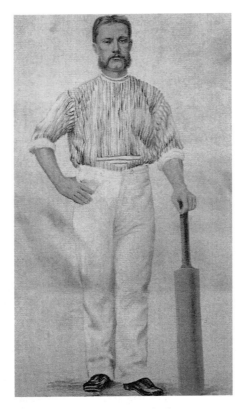

A new Australian sporting hero: Charles Bannerman.

Between 1928 and 1932, the West Indies (1928), New Zealand (1929/30) and India (1932) started to play Test cricket. Each played their first series against England – in fact, only Zimbabwe and Bangladesh, the two most recent additions to the list of Test-playing teams have played their inaugural match against any other side; for these two most recent additions to the Test fraternity, their first match was against India. In 1928/29 England could even afford the luxury of playing Test series simultaneously against both New Zealand and the West Indies; even then the selectors could leave several of their very best players at home and still win both series!

In 1954, newly independent Pakistan, recently partitioned from India, played its first Tests; previously its players had played for India. Pakistan showed how things were changing in world cricket by making an immediate impact, winning only their fourth Test, to share their inaugural series with England 1-1.

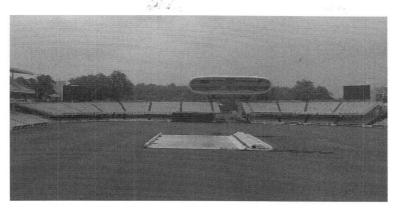

Lords cricket ground on a wet August Saturday in 2008, photographed from the players' balcony of the Home dressing room in the Lords pavilion, with the magnificent media centre perched above the Nursery End stands. Photo: Mark Kidger.

Pakistan's immediate success was a sign that the traditional powers of the game – England and Australia – would not be able to count on continued domination. There were already signs in the 1950s that other sides were starting to challenge them too. By the early 1960s the West Indies had developed to such an extent that their supremely talented side could inflict two severe series defeats on an England side unused to such treatment and also inflict a first series defeat on Australia. South Africa, in their last two series before isolation, got some measure of revenge for past humiliations by inflicting heavy defeats on Australia and India started to show glimpses of the rare talent that was to come.

After several times threatening to end the Anglo-Australian hegemony, by 1977, it was clear that West Indian cricket was now the strongest in the world. Led by a battery of pace bowlers that inspired fear in many batsmen and respect even from the bravest, side after side was ruthlessly gunned down. Test series against the West Indians became exercises in survival, normally terminating in ruthless humiliation for cricket's former powers. By the late 1980s the old order had been shaken so thoroughly, that it could be argued that England and Australia were the weakest of the Test teams apart from new boys Sri Lanka. It was an exciting time for cricket. For example, New Zealand destroyed an Australian team that had previously not even considered them worthy opponents. Fans around the world could see great fast bowlers like Richard Hadlee, Imran Khan and Kapil Dev and mind-blowing spin bowlers such as Abdul Qadir appear from unexpected quarters. One of my greatest memories was of watching a World Cup match at The Oval between perhaps the two finest all-round line-ups of a generation in Pakistan and the West Indies. As a neutral, sat in the crowd, surrounded by the supporters of both sides, watching a titanic struggle on the pitch was an experience that I will never forget. England could not compete with such talent, but it was a joy to watch. However embarrassing it was to see England being destroyed by the West Indians, you could not see it without feeling a sense of awe. As great players like Garry Sobers, Rohan Kahnai, Alvin Kallicharran, Clive Lloyd, Gordon Greenidge, Keith Boyce and Vanburn Holder retired, they seemed to be replaced by an endless stream of players who were even better. Bowlers of the class of Wayne Daniel and Winston Davies, who would have been strong candidates for the All-Time Best XI for any other side, could hardly get a game for the all-conquering West Indians.

By the 1990s cricket had never been so diversified. Although the West Indians were no longer quite so dominant and Australia was back on the rise, the truth was that you could no longer look at any opponent as a soft option and a chance to rest a few players. Any side could beat any side and newcomers Sri Lanka and Zimbabwe rapidly started to embarrass the bigger teams too. The times they were a'changing. That the old order had been turned on its head was confirmed by the ICC's fledgling Test rankings which, in summer 2000, showed England, one of the traditional powers of the game, firmly at the bottom of the pile, below even Zimbabwe.

Although West Indian hegemony was replaced by an Australian dominance into the new millennium, at one time or another South Africa, India, England and Sri Lanka all laid claim to being the nearest challenger to Australia at the start of the 21st Century. Finally, early in 2009, for the first time since the 1970s, a team other than Australia and the West Indies could claim to be the world's best[19]. First South Africa took the crown from Australia and then, later in the year, India took it from South Africa. As of early in 2010, when I am writing these words, the top of the table is sufficiently tight that in the next 12 months any one of India, South Africa, Australia and Sri Lanka could end up leading it[20].

What of the future? The last three sides to be awarded full membership of the International Cricket Council have been Sri Lanka (in 1981/82), Zimbabwe (in 1992) and Bangladesh (2000). Sri Lanka rapidly won the respect of opponents and in just 15 years had become World Cup winners, also enjoying great success, particularly at home, in the Test arena. Zimbabwe also showed themselves to be tough and enjoyed some initial sparkling successes. However, after a series of major internal upheavals that led to a catastrophic decline in results, Zimbabwe withdrew from Test cricket in January 2006 and, at the time of writing, still seems to be at least a year or two from returning. The jury is still out on Bangladesh, A cricket-mad population follows their team with passion but, as yet, at least in Tests has had little to cheer. While Bangladesh has won two series, they were against woefully weakened Zimbabwe and West Indian sides. Against full-strength opposition, despite sometimes getting into promising positions, the result has been consistently disappointing: fifty Tests against Australia, England, India, Pakistan, South Africa and Sri Lanka have resulted in 48 defeats and just 2 draws, both of them with generous assistance from the weather[21].

While Sri Lanka's rise after their entry into Test cricket to being one of the top teams in world cricket was steady and, at times, spectacular, the struggles of Zimbabwe and, especially Bangladesh, have made it more difficult to see further sides joining them at cricket's top table in the near future. The gap between the best and the weakest Tests sides is already a gaping chasm and no one wants to see it grow larger still. When Bangladesh played their first series in Australia in 2003, there was a well-publicised boast that Australia would win one of the Tests in a single day; this was not just Australian bombast, unfortunately, there was a genuine feeling that such a feat might even be possible[22].

With a desire and need to expand the game to new audiences, a proposal was made to expand the number of Test-playing sides to sixteen by promoting the top Associate Members to a Second Division of Test cricket, but this was not well received. No one wanted to add more costly mismatches to an already overcrowded schedule. However, a less ambitious variant on the scheme has seen the establishment of the Intercontinental Cup. In its 2009/10 incarnation, the best six Associate Members of the ICC and Zimbabwe play each other, with the top two teams in the league table playing the Final for a considerable cash prize. The Intercontinental Cup is, effectively, Test cricket's second division, although without the offer of promotion to full Test status for the winners.

Who might become, one day, the eleventh full member of the ICC and graduate to playing Tests? For a time Kenya made a strong case for promotion, but their playing standards have dropped so far in the last few years that they are no longer a serious candidate. In 2009, after dominating Associate cricket for several years, Ireland made a formal enquiry as to their route map for Full Membership, but a small player pool and doubts about infrastructure seem likely to be severe handicaps for them.

For the romantic, one possible candidate for fast-track promotion is Afghanistan. Initially placed in Division Five of the nascent World Cricket League and forced to play their home matches in Dubai, Afghanistan have risen at an astonishing rate. By getting successive promotions to Division One of the World Cricket League, finishing their inaugural campaign just a fraction away from World Cup qualification, earning One Day International status and then, just a few weeks ago, as I write these words in March 2010, winning the qualifying tournament for the 2010 Twenty20 World Cup in the Caribbean and standing top of the Intercontinental Cup table, they have become the great success story of world cricket. Few would bet against them having further success and, possibly, playing Test cricket within ten years and fewer still would begrudge them the accolade should they attain it.

But will Test cricket even exist for Afghanistan to aspire to play in ten years time? Some people are sceptical. Attendances at Test matches are dropping in most countries and interest is falling. Only in England and Australia are full houses more or less guaranteed and Test series, even against less fashionable opponents, profitable concerns. In response to declining interest in limited overs cricket in England, the England and Wales Cricket Board (ECB) proposed a new format of the game, which they called Twenty20. Each side

A view of the hallowed Lords turf from the Test Match Special radio commentary box. The historic Lords pavilion and its Long Room are at the far end of the ground. In the summariser's chair, my father, Ronald Kidger, who was a regular visitor to the ground to watch Middlesex play in the late 1940s and '50s when working for de Havilland in nearby Hatfield. Photo: Mark Kidger.

would have 20 overs to bat, with everything designed to speed-up play and make it all-action to appeal to a family audience and make it convenient for TV by lasting little more than two hours. The game was an immediate success in England and taken up rapidly in Australia and the Caribbean where a regional competition set up by Sir Allan Stanford became an instant success[23]. However, India was deeply suspicious of the new competition, fearing that it would reduce their lucrative income from 50-over cricket. This opposition ended after the success of the inaugural World T20 competition – won by India – and of the unofficial Indian Cricket League. The BCCI launched its Indian Premier League (IPL) to great fanfare in 2008. The IPL has been such a success that there are fears for the more traditional forms of the game. Players can earn such huge sums playing a few weeks of T20 cricket that it has become lucrative to consider globetrotting through the year from franchise to franchise playing just T20. Could international cricket survive if the top players no longer play for their country? Some sceptics suspect that T20 cricket will flourish briefly before overkill sets in: already there are signs that interest is waning in England although going from strength to strength in India.

Whatever the future of cricket brings though, it is certain to bring great changes. There will though always be a niche for the longer form of the game, although this may need to adapt to survive in a market where the demands for instant gratification grow by the year. A reduction of Tests to four days and the introduction of day-night matches seem now to be almost inevitable, changing fundamentally the format of Tests matches for the first time in over a century of history.

Endnotes

[1]   The Belgian Embassy in London has a note in its Web Page that draws attention to the Belgian claim to have invented cricket:

http://www.diplomatie.be/london/default.asp?id=38&ACT=5&content=147&mnu=38

Although organised cricket has been played in Belgium since at least 1866 and the Belgian national team played its first match in 1905, against The Netherlands, their recent tradition in the game is very limited. Belgium became an Affiliate Member – the lowest level – of the International Cricket Council (ICC) in 1991 and rose to Associate status in 2005. They play in the Third Division of the European Cricket Championship (http://en.wikipedia.org/wiki/Belgium_national_cricket_team) and are not involved in any of the seven divisions of the World Cricket League. In 2009, Belgium placed 4th of the six teams in the Championship, behind Israel, the Isle of Man and Spain, but ahead of Portugal and relegated Malta. If Belgium was the birthplace of cricket, they have certainly not developed their prowess at the game in the last 5 centuries.

The Belgian Cricket Federation's Web Site is:

http://www.icc-europe.org/BELGIUM/index.shtml

[2]   http://en.wikipedia.org/wiki/Talk:History_of_cricket_to_1725

[3]   Translation: it may be true, but believe it at your peril! This phrase is habitually used by scientists to say that something is possible, but that does not mean that they actually believe it themselves. While "creckett" is phonetically almost identical to the word "cricket", "creag" requires a larger leap of the imagination as only the first syllable is close to the modern word. What expenses might be incurred by the young prince in playing a primitive form of cricket that King Edward would have to reimburse are hard to imagine, unless the prince was a big-hitting batsman who regularly smashed windows! Again, it is fair to say that doubts have been raised that the reference is even to Edward Prince of Wales.

[4]   Re-reading these words I am struck by the similarities of this form of cricket to the concept of the IPL and the way that the innovation brought cricket to a whole new public.

[5]   At the same time, the MCC relinquished its control of the domestic game in the United Kingdom to the England and Wales Cricket Board. However, the MCC holds the copyright to the Laws of Cricket and is thus, *de facto*, their custodian. Despite this, it remains to be

seen how long this situation will last, particularly as the ICC moved its base away from Lords in 2005, thus breaking its historical link with the MCC and its emblematic ground and increasingly wishes to distance itself from the historical domination of its founder.

[6] Depending on your age and your national allegiance, you would either see this as a long-overdue reform of an anachronistic piece of colonial history, or a slightly sad end to one of cricket's quainter, but harmless traditions. Of course, cricket has based its appeal to a large degree on tradition. For the 2010 English season the MCC brand has been added to the names of the sides that were previously called "University Centres of Cricketing Excellence" (Cambridge, Durham, Loughborough and Oxford UCCEs played First Class matches). These sides are now known as MCC Universities and play official MCC University matches.

[7] An example of this is the game between England and a 33 of Norfolk, played on July 17th, 1797. Despite being played for a wager of five hundred guineas, meaning that the players had plenty of incentive to play hard, the 33 of Norfolk mustered just 50 and 81 for the loss of 62 wickets, with just one batsman, who batted at 28 (!!) in the second innings, reaching double figures. No less than 36 Norfolk batsmen registered ducks in the match.

[8] Century Publishing Ltd, Portland House, London. ISBN 0 7126 0939 3.

[9] One of the great curiosities of cricket is that it has traditionally been played between counties, provinces, or regions rather than towns. With the success of the city franchises of the Indian Premier League (IPL), the idea of re-branding sides has come up once again. Would a London v Manchester fixture be more attractive and marketable than Middlesex v Lancashire?

The USA v Canada match was played to a finish by two representative teams of acceptable strength, on even terms, thus satisfying the basic criteria to be considered a Test match.

[10] Baseball has huge advantages over cricket in this sense as the ball does not need to be pitched – thus there is no need for reliable bounce, so even very rough, stony ground is fine – nor is the ball hit mainly along the ground as it is in cricket, thus obviating the need for very short grass even in the outfield.

[11] With the advent of the World Cricket League and regular matches with First Class status for the top sides in the Intercontinental Cup, there is the prospect that the USA v Canada fixture could, at some point in the future, acquire First Class status.

[12] The only other 11 v 11 match was a two day game v New South Wales in January which Lillywhite's XI drew, forcing New South Wales to follow on 188 behind and having them 140-6 when time ran out. The scorecard is available here: http://static.cricinfo. com/db/ARCHIVE/1870S/1876-77/ENG_IN_AUS/LILLYWHITE-XI_NSW_15-16JAN1877.html. Given the events of later in the tour it is interesting to see that Charles Bannerman could only score 2 and 32, although this latter score was important in helping his side to hang on for the draw.

[13] Lillywhite, conscious of reducing costs, took only 12 players to Australia and New Zealand. Of them, the wicket keeper, Ted Pooley, remained in a New Zealand jail when the rest of the team returned to Australia, having foolishly got himself embroiled in a betting scandal. Harry Jupp, his deputy, had severe conjunctivitis and could not keep wicket, but was obliged to play as a batsman and Tom Armitage, who suffered the indignity of dropping Bannerman in single figures, was reportedly still so weak from sea-sickness that he could barely stand before the start of the match.

[14] Bannerman retired hurt at 240/7 when he split the index finger on his right hand after being hit by a lobbed underam ball from George Ulyett. After his dismissal, Australia lost their remaining 13 wickets in the match for just 109 runs, which says a lot about Bannerman's influence on the result. Apart from Jupp, opening England's first innings, who made 63, no other batsman in the match reached even 40, You can make a strong argument for Charles Bannerman's innings, which was the first ever in Test cricket, also being the greatest ever in Test cricket. It still remains the highest percentage of runs from a single batsman in a completed Test innings (67.3%) and one of the very highest percentages of a completed innings in all First Class cricket. The scorecard for this match is available at: http://www.cricinfo.com/ci/engine/match/62396.html.

In One Day Internationals, the record percentage is even higher. In a quite astounding match that I well remember, England seemed to have the mighty West Indians tottering towards defeat until their number 11, Michael Holding, held firm at one end, allowing Sir Vivian Richards to blast 189 not out, out of a total of 272-9 (69.49%) in 55 overs. A demoralised England crumpled to a heavy defeat when they batted.

The highest percentage of all in a First Class match was by Glenn Turner (83.4%) when he scored 141 not out in Worcestershire's total of 169 against Glamorgan at Swansea in 1977. The scorecard for this remarkable match is at: http://www.cricketarchive.com/Worcestershire/Scorecards/37/37233.html.

Charles Bannerman only ever played in 3 Tests, the two against James Lillywhite's XI and the crushing victory at Melbourne in 1879 against Lord Harris's XI. His other five Test innings summed just 74 runs and he never scored another First Class century in his 44 match career. Bannerman eventually retired from cricket in 1887, before becoming an umpire and standing in 12 Tests between 1887 and 1901.

[15]   One suspects that, in retrospect, the English were more reluctant than the Australians, to include that first match played in Melbourne in 1877.

[16]   The Ashes trophy is tiny and, if truth be told, a little lop-sided. On the day that I visited the Lords Museum though, its somewhat unimpressive aspect did not stop a large crowd of visitors, mostly Indians who worship the game, looking awe-struck at it. The oldest sporting trophy in the world needs no gold or silver adornment to inspire respect.

[17]   England supporters of my generation became used to seeing one-sided contests, with Australia dominating but, being an Australian cricketer in the 1880s and '90s must have been equally demoralising. Between the 5th Test of the 1884-85 series in Australia and the end of the 1890 series in England, Australia lost 11 out of 12 matches, including 7 consecutive Tests over 4 series; this domination lasted most of the following decade, albeit to a slightly lesser degree.

[18]   Until the first South African series victory in 1966/67, South Africa's best results in the ten series played, was two drawn series.

[19]   Theoretically, South Africa had briefly become the #1 in 2002 on the old system for calculating the ranking table in which sides received two points for winning a series and one for a draw, irrespective of the strength of the opposition and the length of the series. As all sides were expected to play all sides home and away, in the end the table would be a reflection of relative the strength of sides over a number of years. This though led to a situation whereby South Africa played all their series against Bangladesh and Zimbabwe home and away, picking up 8 points for it, whereas Australia had not played them at all. Despite heavy defeats in the home and away series against Australia, South Africa topped the table thanks to their wins against weaker opponents and thus, officially, was #1. After this the points system was changed such that results against strong opponents are much better rewarded than against weak ones and winning a series 4-0 gains more points than winning 1-0.

A nice feature of the system is that when the disparity between two sides is great, the weaker side may actually score important points even for battling to a narrow defeat. In each series, dependent on the length of the series and the relative strengths of the sides, there is a par result: do better than this and you gain points, do worse and you lose them. When South Africa and England drew their series in Winter 2009, as South Africa were much higher ranked, they needed to win the series by at least two Tests to gain any points at all; when England drew the series, England's gained points and South Africa lost them.

[20] You can find the official tables, updated after every Test series, here:

http://icc-cricket.yahoo.net/match_zone/team_ranking.php.

[21] The two unlucky sides who failed to beat Bangladesh were India in 2007 (two and a half days lost to rain) and New Zealand in Bangladesh in 2008/09 (three days lost to rain).

[22] In fact, although both Tests ended in heavy defeats for Bangladesh, they showed more resilience than expected and the pre-series boasts were loudly recalled as the visitors threatened for a time to make a real game of the 2nd Test.

[23] Stanford's involvement with cricket came to an embarrassing end when he was indicted for fraud and embezzlement. His high-profile partnership with the ECB was unceremoniously dropped and he was stripped of his knighthood. The Stanford affair became a huge embarrassment for the ECB, West Indian cricket and the government of Antigua. There have been suggestions that many important details of the events surrounding the scandal have not been made public.

CHAPTER 2

WHAT ODDS THE TOSS?

'Well, gentlemen, we shall bat.'

W.G. Grace's habitual response to winning the toss

During the early part of 2009 England played seven Tests against the West Indies: five, including the abandoned Test, in the Caribbean and then two in England. Unusually, the England captain, Andrew Strauss won the pre-match toss of the coin every single time. At least, given that the host captain tosses and the visiting captain calls "heads" or "tails", Andrew Strauss won all five tosses in the Caribbean and Chris Gayle lost both in England.

The toss before the 1st Test between England and South Africa at the Wanderers Ground in 2009. As is traditional, the home captain – Graeme Smith, here – tosses the coin and the visiting captain – in this case, Andrew Strauss – calls. Andrew Strauss won, as has been his habit since becoming captain of England. Is this evidence of skulduggery? No, it is simple Poisson statistics.

The side winning the toss in cricket has an advantage that is not present in other sports. In tennis winning the toss gives the player the choice of serving first or of choosing the side of the court. In football or rugby the choice is usually of end and is only decisive if there is a strong wind blowing. In American football you decide to kick or to receive. None is really a match-changing advantage except in wholly unusual circumstances. In cricket though, the winning captain can choose whether to bat or to bowl. Winning the toss allows the captain to gain a major strategic advantage over his opponent, allowing him to decide how best to deploy his resources. If the weather and the pitch are good, a captain will almost invariably choose to bat in the hope of building a massive score and putting the opposition under pressure. If the weather is threatening, or the pitch suspect, a captain with a strong hand of bowlers will often choose to invite the opposition to bat in the hope of dismissing them quickly and then batting in easier conditions afterwards.

Winning the toss is often seen as significantly increasing the chances of victory. When a one side wins the toss in every Test of a series, that side is often seen to have held an unfair advantage in the series. A study of results of Test matches since 1900 shows that the side winning the toss has held, on average, a 54-46% match-winning advantage[1]; such an advantage is large enough to be important, as it means winning an additional Test for every twelve matches in which the toss is won.

When a side wins the toss in every match of a series there are invariably calls to abandon the toss and simply alternate advantage. These have, so far, been resisted, in part because it would offer a temptation to be unscrupulous (if you *know* that you will get the choice of batting or bowling, you could tailor the pitch to your side's needs: for example, have a pitch that is excellent for batting initially, before breaking-up and taking a large amount of turn… the side batting second would have little chance of avoiding defeat).

What determines the toss? Is it really unfair to win several times consecutively?

The winning and losing of the toss is determined by probability. In other words: pure, random chance. However, probability can be a tricky thing. Wait long enough and anything, however improbable, will happen eventually.

What is the probability of Andrew Strauss winning seven tosses in a row?

A coin, if fair, will have exactly a 50% probability of coming down heads and a 50% probability of coming down tails. So, whatever the call, Andrew Strauss had a 50% chance of guessing right.

So, what are the chances of guessing right seven times in a row? The probability is simply 0.5 to the power of the number of tosses; in this case, seven, i.e.

$$P = 0.5^7 = 0.0078$$

In other words, by chance it should happen only once every one hundred and twenty eight series of tosses that a captain will win seven in a row. Given that, as of February 2010, the number of Test matches played had reached 1950, we can reasonably expect streaks of seven and even eight consecutive tosses won to have occurred multiple times in the history of Test cricket. There is nothing fundamentally outrageous about a captain winning the toss in five, six, or even seven consecutive matches.

By the same token though, Chris Gayle can consider himself singularly unfortunate: not many people would bet much money on odds longer than 100-1.

So, what was the probability that Andrew Strauss would make the sequence eight successful tosses in a row at the SWALEC Stadium in Cardiff against Australia in July 2009?

The probability of winning eight tosses in a row is

$$P = 0.5^8 = 0.0039$$

So there was just a 1 in 400 chance of that coin coming down the way that Andrew Strauss wanted – in other words, of Ricky Ponting calling wrongly?

Wrong!

The probability of Ricky Ponting calling wrong was still exactly 50%: no more, no less.

Even if Andrew Strauss were to win the toss in twenty consecutive Test matches, the odds of which are approximately one in a million, the probability of winning the toss in the twenty-first Test match would still be exactly 50%.

In fact, Andrew Strauss continued his lucky streak both at the SWALEC Stadium and in the 2nd Test at Lords, before Ricky Ponting finally called correctly in the 3rd Test[2].

Probability will always catch up with you in the end. You can only buck it for so long. Provided that the coin that you use is honest, if you continue tossing the proportion of wins and losses will always approach 50% exponentially, however skewed the results are initially.

Let's take a look at another case of probability. Suppose there a fielder has an extraordinarily safe pair of hands and a 98% chance of taking a catch that is offered – in other words, just one in fifty times, which would be an extraordinary percentage, the chance will be missed – how many catches have to be offered before there is a 50% probability of one of the chances having been missed?

The intuitive answer is 50, but it is wrong.

There is a probability of 0.98 or taking the first catch successfully and then a probability of 0.98 of taking the second and a probability of 0.98 of taking the third. So the probability of taking all three successfully is

$$P = 0.98 \times 0.98 \times 0.98$$

Or

$$P = 0.98^3 = 0.94$$

In other words, the chances of taking three in a row are 0.94 or 94%.

That is pretty good odds. In general, the probability of a successful sequence is:

$$P = 0.98^n$$

Where "n" is the number of consecutive catches held

How many catches do we need before the cumulative probability of success reaches 0.5, or 50%? The number is surprisingly small, n=34. For n > 34 the cumulative probability of always being successful in taking the catch drops below 0.5.

However, even if the cumulative probability of taking more than 34 consecutive catches is below 0.5, over a long sequence, the average number of misses will still be once in every fifty: offer 200 catches and you would expect just four to be missed.

We often say that "luck always evens out in the end". This is a basic tenet of probability. If an event is truly random, the longer the sequence the more likely that luck will even out. Sooner or later, if he is captain for long enough, Andrew Strauss will *lose* seven consecutive tosses and will curse his bad luck (strangely enough though, players rarely complain when probability works in their favour).

How does this work in practice?

Suppose we toss a coin one hundred times. We would expect to get 50 heads and 50 tails, or to call correctly 50 times. We would *not* expect to get a 90-10 result.

We would expect a standard deviation in the results of:

$$\sqrt{N}$$

Where "N" is the number of trials; in this case, coin tosses.

67% of the time we would expect the results to fall within 1 standard deviation of the expected result. For 100 tosses

$$\sqrt{N} = \sqrt{100} = 10$$

So we would expect 67% of the time to be within ±10 of the expected result, i.e. between 40-60 and 60-40. The more we move away from the expected value, the less likely the result.

As a proportion of the number of trials, the standard deviation is:

$$F = \sqrt{N}/N$$

$$= 10/100$$

$$= 0.1$$

In other words, we have a 1 in 3 chance of a 10% deviation from the expected result. With only a small the number of trials, there is a relatively large probability of big deviations from the expected result.

Overall, as a fraction of the number of trials, the standard deviation in the result is

$$F = 1/\sqrt{N}$$

Technically this is known as "shot noise", or Poisson statistics. Poisson statistics appear in many places in science and everyday life from the decay of radioactive atoms to insurance premiums. Actuaries base their risks on Poisson statistics: an insurance company does not know when a particular person will die but, if it insures a million people, it can calculate very accurately what fraction of those people will die in any given year and thus how much money it is likely to have to pay out.

So, if we toss the same coin one million times, we expect the standard deviation to be

$$\sqrt{N} = 1000$$

And, as a fraction of the number of trials, it is:

$$F = 1/\sqrt{1\ 000\ 000} = 0.001$$

In other words, 67% of the time, if the coin is honest, the division of heads and tails will be within 1000 of being exactly 500 000 heads and 500 000 tails. The larger the number of times that we toss the coin, the smaller the deviations from a 50-50 split will be as a fraction of the total.

If we toss a coin one million times and get 600 000 heads and only 400 000 tails, we can be pretty sure that the coin is biased and that it is weighted to favour its falling "heads". The larger the number of trials, the closer we will expect to come proportionally to the expected result and the more suspicious large deviations from random behaviour become.

So, can we suggest on the available evidence that Andrew Strauss is somehow smuggling a trick coin onto the field of play in order to win the toss?

No, we can not because, by the same token, the smaller the number of trials, the larger the expected range of deviation: what scientists call "small number statistics", or getting a misleading result because you did not repeat the test often enough.

After 8 tosses, we expect a 4-4 split, but the standard deviation is

$$\sqrt{8} = 2.8$$

So, less than two thirds of the time we expect the toss to split between 6-2 and 2-6. In fact, one third of the time we would expect less than 1.2 or more than 6.8 tosses out of 8 to be won. Winning the toss six times out of eight is no evidence at all of a biased coin being used! Winning it sixty times out of eighty would though be deeply suspicious: it is not impossible that it could happen, it is just very unlikely that probability will break so heavily in your favour unless it has some help[3].

Why is there such a mystique about the toss in cricket? In part it is because of the ceremony involved: in other sports the toss is anecdotal (it is rarely reported in a football or tennis match because it is usually an irrelevance) or, as in American football, purely ceremonial. In cricket the winning captain has to make a major decision that will be widely reported and commented on and he will be mercilessly criticised for making a wrong decision. What is more, in international cricket, the toss is carried out in front of the TV cameras and the winning captain will immediately have a microphone thrust under his nose by a former player and asked to justify his decision to the watching public. In part it is because a statement of intent is being made. If the captain winning the toss decides to bowl first, is it because he thinks that the conditions favour his bowlers? Or is it because he is scared of what the opposition bowlers might do to his side if he bats first? A captain who shows himself reluctant to expose his side is handing a huge psychological advantage to the opposition. Similarly, a captain who is overly aggressive, but cannot back up his intentions with success on the field, is throwing away the initiative. Cricketers, like most sportsmen, are highly superstitious: starting a match with a defeat, even a symbolic one in the toss, is regarded as a bad omen and a significant disadvantage for the match[4].

Now that we have the toss out of the way, let's get on with the game.

Endnotes

[1]   http://cricketanalysis.com/winning-the-toss-in-test-cricket/

Since the 1960s the advantage of winning the toss has appeared to reduce progressively and, since 1990, the cited study suggests that it has effectively disappeared.

[2]   One England captain of recent times was regarded as being particularly unlucky with the toss, despite changing his call between "heads" and "tails". A wag in the dressing room suggested that as neither call seemed to work, he should have tried calling something else instead. The England fast bowler, Bob Willis, regarded himself as being an unlucky tosser. After a particularly outrageous succession of bad luck in matches against New Zealand, when he actually won a toss, he informed the waiting New Zealand captain, his old friend Geoff Howarth, that he was too shocked to decide whether to bat or to bowl.

[3]   W.G. Grace is reputed to be one of those who "helped" probability a little. He frequently called "the monkey" when the coin was tossed, instead of "heads" or "tails" and would then pick up the coin, irrespective of which way up it had fallen and declare "well gentlemen, we shall bat". A contemporary of Grace was once asked if W.G. ever broke the rules. The answer was "no, he was far too clever for that, but it was wonderful what he could do with them".

[4]   Mike Gatting once explained his strategy to win a critical toss in Perth on the 1986/87 England tour of Australia. "I thought 'tails for Wales' and Perth is in the west of Australia". However the decision was arrived at, it worked!!

CHAPTER 3

LIES, DAMN LIES AND AVERAGES

'After all, facts are facts, and although we may quote one to another with a chuckle the words of the Wise Statesman, "Lies - damn lies - and statistics," still there are some easy figures the simplest must understand, and the astutest cannot wriggle out of.'

Leonard H. Courtney (1895),
popularised by Mark Twain (1907)

L ove them or hate them, there is no question that statistics are part and parcel of the game of cricket. However, to paraphrase the immortal Sir Humphrey Appleby in the BBC series, "Yes, Prime Minister" sometimes that the average, probably the most used statistic in cricket, is asked to "carry a greater burden than a simple number can bear". In other words, statistics are as frequently misused as used in cricket.

Cricketers, like baseball and basketball players, live and die by their averages. Although statistics are an ever-present part of most sports and great efforts are made to record with the greatest accuracy how many greens Tiger Woods has hit in regulation through his career, or what Rafael Nadal's first serve percentage is, or how many goals David Beckham has scored in his games for England, in the end, if you win a tournament, it does not matter how many greens you miss, how many putts you have taken, or how many double faults you have served. In cricket though, no matter how your team does, if your average is poor, it is a career-ending number. No batsman can "wriggle out of" averaging under 30 for very long. No bowler can hide for long if he averages over 40. People use averages and throw around numbers to discuss the worth or otherwise of a player; few people though understand the limitations and proper use of averages. Like all statistics, badly used they are at best misleading and, at worst, completely useless.

When Leonard Courtney used the famous phrase "lies, damn lies and statistics" in a speech in New York in 1895 he knew what he was talking about because two years later he would become President of the Royal Statistical Society. However, it was Mark Twain who popularised his words in 1907 in "Chapters from my autobiography". Properly used, statistics are revealing and document inherent truths. Many times though they are not properly used.

Let's take a look at one particularly famous example of abuse of statistics. Because it is so famous, it is frequently held up as an example of how meaningless cricket statistics are.

It is generally reckoned that averaging 100 in an English cricket season is a truly exceptional achievement that is worthy of only the very greatest players. Let's look at the players who have averaged 100 over an English season:

Player	Year	Average
Don Bradman	1938	115.66
Damien Martyn	2001	104.66
Mark Ramprakhash	2006	103.54
Geoff Boycott	1979	102.53
Bill Johnston	1953	102.00
Mark Ramprakhash	2007	101.30
Graham Gooch	1990	100.70
Geoff Boycott	1971	100.12

As can be seen from the table, this particular feat has been accomplished just eight times, six of them since 1970, showing that what was a very rare feat indeed is now becoming relatively commonplace. It is no great surprise to see the name of Don Bradman there – in fact he just missed out in 1930[1] – and not too many people will be amazed to see Geoff Boycott's name given his fame to be, at his best, well nigh undismssable, but was Damien Martyn's performance in 2001 really superior to Don Bradman's near miss in 1930? The fact is that in 1930, Don Bradman sustained his remarkable form through no less than 27 matches and, in 1938, through 20 while, in 2001, Damien Martyn played just 8. Would Martyn have sustained such a remarkable run of form if he had played three times as many matches? One suspects not, but his average still suggests, rightly or wrongly, that his was the second greatest batting performance ever in an English season[2].

The performance though that is the classical indictment of averages is that of Bill Johnston, in fifth place in the above table. He played 40 Tests for Australia between 1947 and 1955 with figures that any player would be proud of... as a bowler. In fact, Bill Johnson had no pretensions whatsoever to be a batsman, passing 20 just three times in his Test career and never reaching 30. Almost always he batted number 11 and when he did not, he played at 10. So how did Bill Johnson get to manage the hallowed average of over 100 for a full season in 1953?

In fact, there was some collusion involved. During the Test series he played in three matches, scoring 22 runs in his 6 innings, with a top score of 9. However, crucially, unlike so many of his team mates, neither Alec Bedser nor any of the other English bowlers could get him out: in fact, he was not out in all 6 innings in that famous series. Over the entire tour he batted 17 times and was not out on 16 occasions[3], scoring just 102 runs, more than half of them in just two innings. In his last two matches, once his average had passed 100, he was carefully protected by both his own side and the opposition. In the very last match of the tour Jack Fingleton recounts in "The Ashes Crown the Year" how Johnstone's captain, Lindsey Hassett, gave him a note to present to Norman Yardley, the opposing captain, when he came out to bat. The note read:

Dear Norman

This will serve to introduce you to one Bill Johnson. His average is over a hundred. Look after him.

Lindsey

In fact, there was no need to worry. In the previous match Hassett had declared the innings closed as soon as Johnson scored his 4 runs and, this time, the other batsman at the crease, Langley, threw his wicket away to protect Johnson from having to face even a single ball. Bill Johnson, a classical batting rabbit had, like the immortal Sir Donald Bradman, managed to average over one hundred over a full English season.

Is it right though to say that Johnstone's *average score* was 102 when his best score on the entire tour was 28 not out (note that what we call "the average" is actually "the average number of runs scored between dismissals")? Not really. In fact, it is, of course, downright misleading. Johnstone never even scored more than 43 even in club cricket for his side, Richmond.

Johnson's feat was a statistical freak, although with the considerable merit that, as very much a non-batsman, he batted so many times that summer without getting out on more occasions. It does show just how misleading an average can be though if presented as the be all and end all of performance.

So, let us look a little more closely at statistics and averages and their meaning. Can we do better than classical averages as a measure of performance?

BASIC STATISTICAL PARAMETERS

Statistics is the science of how groups of numbers behave. We assume in statistics are representative: in other words, even if we change the exact set of numbers that we use, as for the coin toss that we have previously looked at, we still get a number that is representative of all possible numbers.

For the population at large, the most familiar representation of this phenomenon is the Opinion Poll: you ask one thousand or two thousand citizens who they intend to vote for and then assume that those people are going to be exactly representative of the almost sixty million population of the United Kingdom, or the nearly three hundred million population of the United States of America and so say what percentage each candidate or party will get in the up-coming election[4].

Statisticians use three basic numbers to define the behaviour of a group of numbers. These are the Mean (often called "the average"), the Median and the Mode.

Let's look at Don Bradman's famous Test career and see how it shapes up using these different statistics. His scores in his 52 Tests were:

18, 1, 79, 112, 40, 58, 123, 37*, 8, 131, 254, 1, 334, 14, 232, 4, 25,
223, 152, 43, 0, 226, 112, 2, 167, 299*, 0, 103*, 8, 66, 76, 24, 48,
71, 29, 25, 36, 13, 30, 304, 244, 77, 38, 0, 0, 82, 13, 270, 26, 212,
169, 51, 144*, 18, 102*, 103, 16, 187, 234, 79, 49, 0, 56*, 12, 63,
185, 13, 132, 127*, 201, 57*, 138, 0, 38, 89, 7, 30*, 33, 173*, 0

For the famous final figures of:

Matches	Innings	Not out	Runs	Highest score	Average
52	80	10	6996	334	99.94

The Mode:

This is the simplest of all statistical numbers. It is the most common score in a list of numbers and so is regarded as being the most representative number of the list. So, if we take the score that Don Bradman obtained most often we would have a representative idea of how good he was, agreed?

Well, let's see how he does.

We find that he was out for a duck (i.e 0) seven times in his Test career. He scored 13, three times and no other score more than twice (remarkably, he scored both 103 and 112 on two occasions and, at the other end of the scale, also scored 1 twice). So, Don Bradman's most representative score was 0: is this a fair reflection of his batting? Not really! On 70 occasions he scored more than 0 runs, so this extreme number is in no way representative of his performances as a whole.

In fact, if we look at almost any player who played more than a handful of games, we will discover that the Mode of his scores is almost always 0.

One of the basic tenets of statistics is that there will always be exceptions to any rule because that is their nature but, in the long term, these exceptions always even out and tend to disappear. In this case perhaps the most famous exception is England's former wicket keeper, Geraint Jones who, remarkably, played 33 Test matches without a single duck before, ingloriously, recording two ducks in his final match. Geraint Jones' Mode is 4, which he scored 4 times in his career: he also scored 2 and 3 more times than he scored 0.

Of current players, England's James Anderson had, up to the start of the 2009 season, played 35 Tests without ever being OUT for a duck. However, no less than 10 times up to then he had been 0 not out and, in his case, having a Mode of 0 would probably be a fair reflection of his ability as a batsman (although Jimmy Anderson himself might dispute that), especially when we consider that he has also scored just 1 on another 7 occasions, so that in all, in 20 of his 39 innings up to then he never reached 2.

Of course, this sequence of – relative – success as a batsman ended in summer 2009, after I had finished the first draft of this chapter. Against Australia, in the 5th Test at The Oval, Jimmy Anderson finally registered his first duck in his 38th Test. Having finally known batting failure, he apparently developed a taste for it, as his second duck followed just five matches later, in South Africa.

These examples show how the Mode is no use to us whatsoever when we want to know how good a player's record is. Most players, however good, will have a Mode of zero or, even if not exactly zero, very close to zero. The longer a player's career is, the more certain that the Mode of that player's scores will be zero because the distribution of any player's scores, however brilliant that player is, will always be skewed towards zero. You can only buck probability for so long so, any player who plays for long enough will always end up having zero as the Mode of his or her scores.

The Median:

The Median is the middle number in a list, ordered by size. This often makes it a good number to use for scientists as it is not affected by a few crazy values. Let's suppose a player has the following series of scores where, as is usual in cricket, an asterisk stands for not out:

0*, 2, 8, 0, 5, 15*, 65*, 6, 4, 20*

That is a total of 125 runs for 6 times out, giving a quite respectable average of 20.83. However, 9 of the 10 innings are of less than 20 and 7 of the 10 do not even reach 10. Is an average of almost 21 a fair refection of the player's ability? We would think that a player who does not normally reach double figures is not exactly a good batsman and it could be that, like Bill Johnstone, he just had one or two lucky innings that make his average look good despite his ability being nothing like as great as the raw figures suggest.

Suppose we order is scores from smallest to largest. We get:

0, 0*, 2, 4, 5, 6, 8, 15*, 20*, 65*

The Median is the middle score. In this case, in a list of 10 it is the average of the 5th and 6th highest scores. Thus, in this case, the Median score is 5.5.

The Median means that half the player's scores will be higher and half will be lower, so it is a fair reflection of ability. However, it cannot take into account not out scores – a score of 20 and a score of 20 not out count exactly the same – so some people would say that the batsman is unfairly punished by this method of calculation. However, let's see how Don Bradman fares using the Median of his scores.

His Median score turns out to be 56.5, much lower than his average, but still quite exceptional. In part the difference between the Median and the Mean is the fact that his ten not out innings summed a quite astonishing 1128 runs, something that boosts his average quite significantly.

In 42 of his 80 innings (53%) Don Bradman passed 50 – hence the Median score higher than 50, something that is unique in all Test cricket.

On 29 of these 42 occasions (69%) he went on to score a century.

Of the 29 occasions that he reached a century, on 18 (62%), he went on to pass 150, on 12 (41%) he passed 200 and on 5 (17%) he passed 250[5].

What then is the difficulty with the Median? The biggest problem is that to get a sensible result you need to have many values when making the calculation and for them to be distributed in such a way that they are reasonably closely spaced. Consider the case of a player with the following series of scores:

21, 5, 63, 18, 96, 88

We order the scores we get:

5, 18, 21, 63, 88, 96

In this case, the Median is the average of 21 and 63: in other words, 42, even though there is no score even close to 42 in the sequence.

Does it make any sense to say that the Median score is 42? Most people would argue, reasonably enough, that it does not. Only for players with long careers, consisting of tens of matches, is the Median really meaningful. In general though, when a player has played enough matches, the Median score is an extremely reliable indicator of ability. Its biggest inconvenience is that it has large *granularity*. The Median of a set of scores can only ever be an integer (i.e. a whole number), or half way between two integers. If we take the Median as our criterion many players will end up with an identical value. Looked at another way though, this may actually be a fair reflection of the real state of affairs: can we really say that there is a significant difference in ability between a player who has had 100 innings and averages 51.56 and another who has played 63 innings and averages 52.01? Probably the Median score for both will turn out to be identical suggesting, correctly, that there genuinely is little to choose between the two. For those who want to tune the average to the second decimal place and say that one player was a fraction better than another, the Median will be unsatisfying in that it will rate many players as being effectively equally good.

However, there is another proviso in the use of the Median, which we will see when looking further at the Mean.

The Mean:

The Mean is often called the Average. Technically, the Mean is

$$\bar{x}_{scores} = (\Sigma_i\, x_i)/n$$

In other words, it is the sum of all the values divide by the number of values summed. In this case, the total number of runs scored divided by the total number of innings.

In cricket though, traditionally it is not the Mean score that is calculated, but the Mean number of runs between dismissals. In other words:

$$\bar{x}_{dismissals} = (\Sigma_i\, x_i)/(n-k)$$

Where

n is the total number of innings

x_i are the individual scores in each innings

And

k is the number of not out innings.

So, unless the batsman has no not out innings in his career, which would be extremely unusual

$$\bar{x}_{dismisals} > \bar{x}_{scores}$$

The average number of runs between dismissals must be larger than the batsman's average score and thus, his true value to his team[6].

The larger the number of not out innings, the greater the difference between the two values.

For Don Bradman

$$x_i = 6996$$
$$n = 80$$
$$k = 10$$

So

$$\bar{x}_{dismisals} = 99.94$$

and

$$\overline{x}_{scores} = 87.45$$

One thing that we notice is that the Mean score of Don Bradman is about 50% large than the Median score. We get an almost identical result for Sachin Tendulkar: his Median score is 31 – that is, half the time he comes out to bat he will score more than 31 and half the time less – but his Mean score is 44.20 (remember that his average is a much higher 54.58).

In statistics the Mean has significance when the numbers used to calculate it follow a Normal, or Poisson distribution, with a lot of values close to the Mean and increasingly fewer as we go away from the Mean, either to higher or to lower values. However, a cricketer's scores almost never follow a normal distribution; in fact, they are massively skewed with, as we have seen, with the largest number of values at one extreme – small scores – and a few very large scores at the other extreme. However, the Mean is very sensitive to having a small number of wildly discrepant values included in the calculation. When used in science, it is normal to exclude such values as unrepresentative outliers and not use them in the calculation of the Mean. So, in reality the Mean is not a good measure to use in cricket because the results can be highly misleading: one big score, especially if it is not out, can make a mediocre series of performances look good.

Consider the case of the English batsman, Robert Key. He has played 15 Tests, enough to have quite reliable statistics. His innings, in order, are:

17, 30, 34, 1, 1, 47, 23, 0, 52, 3, 14, 18, 4, 221, 15, 29, 4, 6, 93*, 10, 0, 41, 83, 19, 1, 9

This gives him 775 runs in 26 innings, with one of them not out, at an average of 31.0: not an outstanding figure, but at face value a fairly respectable one.

However, we notice that 314 of his runs (40.2%) came in just two innings, one of them not out. Remove these outlying values and his average plummets to 19.2[7], a value that is well below the standards expected of a specialist batsman in Test cricket. If we exclude the top three scores, we find something interesting. The Mean drops to 15.8 (fans of Rob Key may think that doing this is a "Mean" trick to besmirch his name), almost exactly the same as his Median score of 15 for all his innings. In fact, it is an illustration of something very curious. If the numbers that we are using follow a normal distribution, in other words the famous bell-shaped curve of chance, tending to a certain typical value and with fewer as we move away from this most common value, the Mean and the Median will give the same result – both will find the central value of the distribution. If the distribution is skewed to low values, the Median will be smaller than the Mean. If, in contrast, the distribution is skewed to large values, we will get a Median that is larger than the Mean.

What we find with Rob Key's performances is that he has an unusually random distribution of scores around a typical value if we discount a very small number of outstanding innings. In contrast, for Don Bradman, or for Sachin Tendulkar, the distribution is intensely skewed to small values, with a very long tail to high values rather than there being just a few "outliers".

What does this say about Rob Key's record in cricket terms compared to other players?

A batsman is most vulnerable when he has just come in to bat and is not warmed up with his eye in and the adrenaline flowing. Any bowler worth his salt knew that when Don Bradman came in to bat, if you did not get him out quickly you would spend a long time trying to correct that mistake. Commit an error and drop Sachin Tendulkar early in his innings and the result would, most likely, be that you do not get a second chance before he has scored a lot of runs. When England were being put to the sword by the great Viv Richards[8] in Test matches against the West Indies in the Caribbean in 1981, Geoff Boycott records[9] how Bob Willis, the England vice-captain made the revolutionary suggestion that rather than try to get Viv Richards out, extra defensive fielders should be placed to stop him scoring easy runs and getting his adrenaline flowing and thus frustrate him into an error before he started to hammer the ball everywhere. In the end England did start to use this tactic.

Over that Test series his scores were:

29, 0, 182*, 114, 15

So, you can judge for yourself how successful the tactic of frustrating him early in his innings was.

Notice that even though is average was exactly 85, three of his five scores were lower than 30. Once Viv's adrenaline was flowing though and he got started scoring, it was like trying to stop a runaway train!

Either way that you try to do it, you aim to get a batsman out before he has got a start and while he is still vulnerable. This is the reason why a batsman's scores are skewed to low values. Once a batsman has got a start (usually reckoned to be 15-20 runs), their timing is normally good enough that only a mistake, a loss of concentration, or the unexpected will dismiss them. From then on it is a matter of the batsman's own ability to keep going that determines how many runs he will score.

So, you would normally expect the distribution of a batsman's scores to follow not a Normal Distribution, but an Exponential Decline with a long tail; the better the batsman, the longer the tail of the distribution will be because, once past a certain score it becomes increasingly difficult to dismiss them, normally, until exhaustion takes over and a fatal error is made. This is what you see for Bradman or Tendulkar. In contrast, Rob Key's distribution of scores is much closer to a Normal Distribution. Even though 16 times in 26 innings he has got a start and got into double figures, only rarely has he managed to cash-in on that start. Effectively, his figures show that until he has got past 50 he remains just as easy to dismiss at any time, which is normally a reflection of poor concentration. Don Bradman, on the other hand, has an enormous skew in his figures. Nineteen times he was dismissed for less than 15; once he got past 15 though, on 69% of occasions he would score at least 50... get him out fast, or expect to chase leather in the field for a long time[10]!

SERIES AVERAGES

During Test or One Day International series, tables of players' averages are published to compare performances. Press, players, administrators and fans will pore over these tables in an attempt to decide whether a particular player has pulled his weight in the side and deserves to continue to be selected. Just how reliable are the figures?

These days there is an increasing tendency to play short series of Tests. Few series are now played over the traditional length of five or even six matches. Series made of two, three, or occasionally four matches are the norm, with only a few traditional contests being played over five matches. What effect does this have on averages, you ask? The answer is quite simple. In an a fairly evenly contested five-Test series a player who was selected for each match might expect to have from eight to ten innings. In contrast, in a two-match series obviously no player can bat more than four times. Averages are statistics and rely on the numbers being used to calculate them being representative of a random sample of that player's scores, based on his talent and form. A scientist will talk contemptuously of something being "small number statistics" – a pretty severe condemnation that some numbers are worthless because there are too few values to be statistically significant and thus we can expect large variations from expected behaviour.

In general, in any physical experiment, even in the most favourable circumstances of well-behaved numbers, statistics based on less than four independent values are completely useless. In cricket, where luck and random factors such as a player receiving a completely unplayable delivery at the start of his innings, or one player batting in poor conditions while another does bats in good light against tired bowlers, even 4 performances is far from being sufficient to make anything other than the broadest judgements (a player with 3 centuries in 4 innings can reasonably be said to be in quite imperious form and a player with three low scores in four innings is obviously struggling a bit – anything more is imposing a greater burden on statistical analysis than the numbers can reasonably be asked to bear).

All in all, the averages in a two-match series are little more than somewhat indicative of ability[11]. By a conjunction of luck (good or bad) and not outs a genuine batting rabbit such as New Zealand's Danny Morrison can have an average in a short series that is higher than a famed batsman such as Steve Waugh. Over a five-Test series it is almost certain though that class will out and that the natural order will be restored. Take performances over a calendar year and, in general, the very best batsmen will have the highest average and a player like Danny Morrison will come out rather badly in comparison. Even so, a player with a high average over a year may have done so thanks to playing against weaker opposition.

In First Class season averages the problem is normally avoided by using a qualification to be included. This is included as a footnote to the tables, e.g. "qualification, 10 completed innings", or "qualification, 10 wickets". This means that any player who has not been dismissed at least 10 times, or taken at least 10 wickets and thus has reasonably representative statistics is excluded from the tables. This though is really only possible in the English season where a side will play a minimum of 16 First Class matches. In other countries the season normally contains too few games to allow this luxury of putting a realistic qualification mark for a player to be included in the season's statistics.

The lesson is that there are no more worthless numbers in cricket than averages for a short series of Tests. Over a series of five matches the averages are usually a good indication of *relative* ability between players, provided that they have all played the same number of matches. Even so, a series of not outs or a "freak" score can skew a player's batting average completely. The larger the number of values that we use to calculate statistics, the more likely they are to be representative of the population at large. In cricketing terms this means smoothing out variations due to form, conditions, the opposition and the many other factors that can influence a cricketer's performance. For this reason the metric that is often used to compare players is the career average.

CAREER AVERAGES

Averages are *relatives*, not *absolutes*. This fact causes more controversy than almost anything else in cricket. The conditions and the opposition will, logically, influence the result. A player who plays mainly in friendly conditions against weak opposition will have an enormous advantage over another who plays mainly in difficult conditions against strong opposition. The former might have a much better average but, in the same conditions against the same strength of opposition the latter may be revealed to be the better player. Because an average is a relative and not an absolute you can only compare them reliably in the same conditions against the same opposition.

One of the classic arguments in modern cricket is the one about which is the greater bowler: Shane Warne, or Muttiah Muralitharan? Both are undeniably great players and both have exceptional records.

Shane Warne:

145 matches, 708 wickets, average 25.41

Muttiah Muralitharan[12]

127 matches, 770 wickets, average 22.18

Comparing the two, Muralitharan has taken more wickets in fewer matches at a better average. So, does that mean that he is the better of the two?

Muralitharan has played 70 of his Tests in Sri Lanka, taking 472 wickets at the remarkable average of 19.36. Sri Lankan pitches are generally reckoned to favour spin bowlers, so his good record at home is not unexpected and gives him a huge boost in his career figures.

Shane Warne's 69 Tests in Australia have produced 319 wickets at a far inferior average of 26.39.

Advantage Muralitharan again? Perhaps, but most pitches in Australia, except at Sydney, are generally reckoned to suit fast bowlers far more than spinners. So again, we are not comparing *like* with *like*.

What about comparing all matches NOT played in each player's home country?

Shane Warne:

76 matches, 389 wickets, average 24.61

Muttiah Muralitharan:

57 matches, 298 wickets, average 26.65

This time Shane Warne has a slight advantage, but is it genuine? Shane Warne never had to bowl against Australia, for most of his career the very best side in the world. In contrast, Muttiah Muralitharan has played 13 matches against Australia with far poorer results than against other opposition (average 36.06, as against an average of 22.18 against all other opposition).

However, it is not just playing against strong opposition that can skew the statistics. Playing a lot against weak opposition can do it too. A regular accusation is that Muttiah Muralitharan has taken a huge number of wickets against Bangladesh and Zimbabwe, the two weakest teams in Test cricket.

Here there is some substance to the suggestion, as Muttiah Muralitharan has taken 176 wickets at an average of just 15.09 in 25 Tests against these weakest of opponents. In contrast, Shane Warne played just 3 Tests against them – 2 against Bangladesh and 1 against Zimbabwe – taking 17 wickets at 25.70. It is also true that Shane Warne has played in a much stronger attack than Muttiah Muralitharan and has had more competition to take wickets from his fellow team members.

You can only compare statistics if they come from the same *population of values*. It is useless to argue from averages alone which bowler is the better of the two when the conditions, the opposition and the strength of ones own team each affect the results.

To get the same population of values, or at least a better approximation to it, we should take only performances against the same strong opposition (India, England, South Africa, Pakistan, West Indies and New Zealand) and only for away matches. Applying this test we get the following results:

Shane Warne

62 matches, 308 wickets, average 25.98

Muttiah Muralitharan

41 matches, 231 wickets, average 24.92

When we try to select from the same population of values, the results for the two bowlers are almost identical, with a very small advantage for

Muralitharan. Even so, you could argue that the exact proportion of matches against each opponent is not the same and that may still skew the results. However hard we try, it is impossible to make an absolutely fair comparison, taking results from identical populations of values, even for two players who have had almost simultaneous careers. All that we can conclude is that both are extraordinary players, but we cannot prove positively that one or the other was the better of the two.

Comparing averages for players from different epochs is a totally meaningless exercise: rules have changed (e.g. the lbw law, the size of the wicket); the nature of pitches has changed; the strength of opponents is different (there is far more strength in depth in Test cricket now than there was, for example, in the Golden Age of the 1930s, when almost all the best players in the world were either English or Australian); the bat and the ball have changed; even the way that the game is played has changed completely. There is no way that we can make the populations of values to be compared similar enough to make even a first approximation at a valid comparison. Whatever your metric, you can say with a high degree of confidence that Don Bradman was the best batsman of all time because his statistics are so far in advance of anyone else, before or since, but you cannot extrapolate that to say whether or not Jack Hobbs was a greater batsman than, say, Sanath Jayasuriya.

What stat is that?

Here is a practical example of how using different statistical criteria gives different results giving different information about the relative merits of players. The table below shows the top ten batsmen in averages for the 2008 English First Class season. A minimum criterion of 20 innings has been applied. As usual the averages are ordered in reverse order, from the highest downwards.

We see that Lance Klusner, Northamptonshire's South African all-rounder finished top, with an average of 73.00, well ahead of Ian Trott and Mark Ramprakhash.

Table 1. English First Averages 2008 (Qualification: 20 innings), ordered by decreasing average.

Name	Team(s)	Div.	M	I	NO.	Runs	Highest	Average	Mean	Median	Mode
L Klusener	Northamptonshire	2	14	20	5	1095	202*	73.00	54.75	57	0
IJL Trott	Warwickshire	2	16	25	5	1240	181	62.00	49.60	37	7
MR Ramprakash	Surrey	1	14	23	3	1235	200*	61.75	53.70	29	17
MW Goodwin	Sussex	1	16	25	2	1343	184	58.39	53.72	28	0
CJL Rogers	Derbyshire	2	16	27	3	1372	248*	57.16	50.81	39	20
HD Ackerman	Leicestershire	2	16	26	3	1302	199	56.60	50.07	22	0
JA Rudolph	Yorkshire	1	16	24	1	1292	155	56.17	53.83	44	0
SC Moore	Worcestershire	2	16	30	4	1451	156	55.80	48.37	35	N/A
RS Bopara	England Lions, Essex, MCC	2	15	26	3	1256	150	54.60	48.31	42.5	7
N Pothas	Hampshire	1	14	23	5	963	137*	53.50	41.87	37	44
Overall				249	34	12549	248*	58.37	50.40	37	0

Now we will re-order the table using the different criteria to see how it changes (Table 2). Given that concern is frequently expressed that a surfeit of "not outs" will falsify averages. It is an old saw on the cricket circuit that some players play for their average, making great efforts to add a not out to their scores. In the top ten of the averages we have a range of influence of not outs from Jacques Rudolph's one not out score from twenty-four innings, to the five not outs that Lance Klusner has in just twenty innings. So, let us get rid of the not outs and re-calculate so that instead of finding the average number of runs between dismissals we have the average number of runs scored per visit to the crease.

Lance Klusner is still top, although now only by a small margin. However, his figures are still very impressive with an expectation of 55 runs from him in each visit to the crease. The biggest change though is in the next positions. Unsurprisingly, the batsmen with fewest not outs, Jacques Rudolph and Murray Goodwin leap up the table from 7th and 4th, to 2nd and 3rd respectively and Jonathon Trott drops from 2nd to just 8th: in his case, his average is significantly padded with not outs. An alternative way of seeing what expectation we may have of a particular batsman is to look at the Median score (Table 3). With a minimum of twenty innings the median should start to be highly significant. The higher the Median score, the more consistent are the batsman's contributions to his team.

Table 2. English First Averages 2008 (Qualification: 20 innings), ordered by decreasing Mean score per visit to the crease.

Name	Team(s)	Div.	M	I	NO.	Runs	Highest	Mean
L Klusener	Northamptonshire	2	14	20	5	1095	202*	54.75
JA Rudolph	Yorkshire	1	16	24	1	1292	155	53.83
MW Goodwin	Sussex	1	16	25	2	1343	184	53.72
MR Ramprakash	Surrey	1	14	23	3	1235	200*	53.70
CJL Rogers	Derbyshire	2	16	27	3	1372	248*	50.81
HD Ackerman	Leicestershire	2	16	26	3	1302	199	50.07
IJL Trott	Warwickshire	2	16	25	5	1240	181	49.60
SC Moore	Worcestershire	2	16	30	4	1451	156	48.37
RS Bopara	England Lions, Essex, MCC	2	15	26	3	1256	150	48.31
N Pothas	Hampshire	1	14	23	5	963	137*	41.87

Table 3. English First Averages 2008 (Qualification: 20 innings), ordered by decreasing Median score.

Name	Team(s)	Div.	M	I	NO.	Runs	Highest	Median
L Klusener	Northamptonshire	2	14	20	5	1095	202*	57
JA Rudolph	Yorkshire	1	16	24	1	1292	155	44
RS Bopara	England Lions, Essex, MCC	2	15	26	3	1256	150	42.5
CJL Rogers	Derbyshire	2	16	27	3	1372	248*	39
IJL Trott	Warwickshire	2	16	25	5	1240	181	37
N Pothas	Hampshire	1	14	23	5	963	137*	37
SC Moore	Worcestershire	2	16	30	4	1451	156	35
MR Ramprakash	Surrey	1	14	23	3	1235	200*	29
MW Goodwin	Sussex	1	16	25	2	1343	184	28
HD Ackerman	Leicestershire	2	16	26	3	1302	199	22

Here we see just what a remarkable season Lance Klusner had with the bat in his last season of English First Class cricket. He is the only player to have a Median score over 50 and his advantage over any other player is huge. Alert readers will notice that Klusner's Median score of 57 in 2008 is even higher than Sir Donald Bradman's Test Median score: in eleven out of twenty innings in the season Klusner passed fifty and, every single innings in which he passed thirty he went on to fifty. His performances were a model of consistency and reliability, even more so bearing in mind that he batted in the lower middle order where innings could be expected to be cut short by declarations or running out of partners. The numbers show that Lance Klusner was, quite genuinely, the stand-out batsman of that English summer.

Using the median score as the criterion, we see that Jacques Rudolph and Ravi Bopara also show themselves to have been particularly consistent scorers in 2008, each with a Median score over 40. Bopara rises from 9th in the First Class averages to having the third highest Median score and Jacques Rudolph rises from 7th to 2nd.

In contrast, Leicestershire's Hylton Ackerman was particularly inconsistent. An average of 56.60 at the end of the season looks extremely impressive, but a median score of just 22 shows that his high average was based on a small number of very large scores counter-acting the weight of a large number of small scores. No less than 55% of all his runs for the season came in just five of his twenty-six innings. The median shows that when his eye was in and the conditions favourable, he made the opposition pay, but in twelve of his twenty-six innings he did not even reach twenty.

Interestingly, we see that despite an impressive average, the enigmatic Mark Ramprakhash also does poorly on the Median criterion, dropping from 3rd to 8th. A generation of cricket lovers have wondered how it is that Ramprakhash could be so devastating in county cricket first for Middlesex and then for Surrey, but have such a mediocre record for England. Here we have a clue as a staggering 69.5% of his runs in 2008 came from just six of his twenty-three innings. In more than half his innings he failed to reach 30. During his England career it was often suspected that he reached the twenties and then got out far more often than he should. This, in turn, suggests that he takes longer to get well set at the crease – perhaps due to a slight vulnerability that bowlers can exploit early in his innings – than a truly great batsman would.

There is a widespread feeling that, although bowling averages tend to be quite informative, batting averages would be graded as "could do better" as they have many caveats and can easily be falsified, especially by one big not out innings. No alternative to the classic average has ever caught on, although it has frequently been suggested that eliminating the compensation for not outs so that averages are genuinely a measure of runs per visit to the crease would eliminate some of the worst abuses of the average as currently used. In contrast, the median is a reflection of a batsman's consistency of scoring. If we see a batsman with a high average, but low median score, we know that the average has been inflated by a combination of not outs and occasional big scores. The closer that the median score is to the average, the more consistently good or bad the player's scores are shown to be. A player with a high average and low median may not be a bad thing for a side as it leads to the expectation of occasional big, match-winning contributions. A player with similar average and median will grind out the runs consistently match after match, but will be less inclined to make the big scores that turn a game on its head.

What about using the standard deviation or the skew to get additional information on batting performances?

Calculation of the standard deviation of the mean of a set of numbers is one of the most basic statistical analytical methods. Usually written as σ, the standard deviation is defined as:

$$\sigma = \sqrt{\frac{\Sigma(X - M)^2}{n-1}}$$

Where

    X is the mean value

    M are individual values

And

    n the number of values used

The standard deviation tells us within what range above and below the mean we will find 67% of all values. So, in theory, 67% of a batsman's scores would be within 1σ of the mean. Unfortunately though, the standard deviation only has meaning when we have a Normal Distribution of values, or something close to it, around a central score. As we have seen, even Don Bradman did not have that kind of distribution of scores. In fact, any batsman will have something approximating to an exponential distribution of scores with the peak at zero. The standard deviation is totally meaningless as a metric for quantifying batting performances.

How about using the Skew, or skewness, of the distribution of scores? Surely, a great player like Don Bradman will reveal himself by having a long tail to his distribution of scores and thus a large value of the skew? The skew is often defined as:

$$\text{Skewness} = \frac{\Sigma \left(\frac{x - \bar{x}}{\sigma} \right)^3}{n}$$

Where

x are the individual values

$\bar{x}$ is the mean value

σ is the standard deviation

n is the number of values

However, it suffers from the same problem that it assumes an underlying normal distribution, with a small number of extreme values quite unlike a cricketer's real distribution of scores and depends on the standard deviation. A similar problem results from an alternative method, the first and second Pearson skewness coefficients:

$$3 * (\text{mean} - \text{mode})/\sigma$$

And

$$3 * (\text{mean} - \text{median})/\sigma$$

Respectively.

In both cases we need large numbers of scores to get a meaningful result and still rely on the use of a meaningless standard deviation to get it. Indeed, as the modal score for any batsman will tend to zero, the first skewness coefficient simply tends to:

$$3 * \text{mean}/\sigma$$

Thus, unfortunately, neither the standard deviation, nor the skewness give statistically meaningful information that help us to define a batsman's performances. However, when you combine the average with the median you do get far more information about the batsman's playing contribution.

If we wish averages to be more informative and less misleading, adding column for the median score and mean number of runs per visit to the crease, alongside the average would be one simple way to do it. Tradition though suggests that this will never happen.

Endnotes

1   In 1930 he scored 2960 runs in 36 innings, with 6 not outs. Had he scored just 40 more runs in the season, or had one more of his innings been a not out, he would have surpassed the magical 100.

2   Averaging 100 or more in a season is now easier probably largely because a player no longer has to sustain his form through so many matches. If the First Class county season gets reduced even further the feat is likely to become even more common.

3   Bill Johnson's one slip was when he was out for 8 against Hampshire at Southampton, caught and bowled by Victor Cannings on June 6th. A top score of 27 against The Gentlemen on August 27th, in one of the games at the end of the tour, after the Test matches had finished, put him on 98 runs for the season. A wild swing for 4 in his next innings against the South of England on September 3rd, took him to his final total of 102 runs and, with only one dismissal, an average of 102 when the final match of the tour in Scarborough ended on September 11th. Once he reached that magic mark neither team-mates nor opponents really wanted to risk him losing it. This was a part of cricket that was only possible in the past: can anyone imagine Ricky Ponting giving a note to Andrew Strauss asking him not to dismiss one of his players when Australia play England? Somehow I think that such a request would meet with short shrift along with a large dose of scandalised outpourings in the press.

A little known element of this story, given in Gideon Haigh's obituary tribute to Bill Johnstone, is Victor Cannings' own reaction as it became obvious what was happening. Among those who urged Johnston on was Cannings himself, who wrote to him: 'Resist all offers of promotion [in the batting order]. I want to be the only man who got you out.'

4   If only it were that simple! In fact, polling organizations know full well that if you pick one thousand people at random it is not usually going to be representative of the voters in general in an election and so apply a series of corrections to get their final result. For example, if you interview one thousand people coming out of a factory, it is likely that you will get a completely different set of results to if you interview one thousand people leaving an expensive department store. The idea is to choose the people that you ask carefully so that they are as close to representative of the whole population as possible and then apply various checks and corrections (who did you vote for in the last election?) and see if they give the correct, known result: if they do not, you know that there is a bias towards

one party or one candidate in your sample of voters and you "correct" your results to compensate (some voters refuse to answer, some just lie and say that they voted for one party when they really voted for another, some genuinely cannot remember who they voted for last time and give the wrong answer, and some just have no intention of voting anyway but say that they will when asked, so the final correction will always be uncertain and open to interpretation). This is one reason why opinion polls, often disagree with each other widely, rarely are better than fairly good at predicting the results of elections and have often been completely wrong. Each polling organisation has its own way of selecting the people to speak to, has its own set of questions to ask and its own way of correcting the results to get what it hopes is the correct answer.

5 As a percentage of all innings, Don Bradman's conversion rates are barely credible:

	Number	Percentage
50+	42	53
100+	29	36
150+	18	23
200+	12	15
250+	5	6

In general, scoring a century every 10 innings is regarded as a very good mark. Sachin Tendulkar, regarded by many both in India and elsewhere as being one of the greatest batsmen ever, has managed 42 centuries in 261 innings in 156 Test matches, or one every 6.2 innings: a quite remarkable 16% of his innings have been centuries and 95 (36%) have been scores of 50+. If we compare Bradman's percentage of scores higher than each particular landmark with Tendulkar's, we will see just what a phenomenon Bradman was:

	Tendulkar	Bradman
50+	36	53
100+	16	36
150+	7	23
200+	2	15
250+	0	6

Even when compared to a player who many revere almost as a god, Don Bradman's figures stand supreme. Slightly more than once in every seven innings Don Bradman scored 200 or more. In contrast, for Sachin Tendulkar the figure is approximately once in every fifty innings. For comparison Sachin Tendulkar's Median score is 31, as against Don Bradman's 56.5, relatively less in about the same proportion as his average is to Don Bradman's.

6   Consider an extreme example. Suppose you have two batsmen in a side. One bats at number 9 and scores 15*, 8, 10*, 5*, 25*, 12* and 5 for an average of 40; the other scores bats at number 3 and scores 52, 16, 40, 42, 28 and 39 for the lower average of 36.2 – which is more valuable to the team? The former may have an average of 40, but has only scored 80 runs in 7 visits to the crease, while the latter may have a lower average, but has scored no less than 217 runs in one less innings. There is no question that the average, in this case, is no guide to which player is more valuable to his team.

7   By tradition, averages in cricket are always calculated to two places of decimals. However, except for players with very long careers the second place of decimals is not significant and even in such cases is not informative as the second decimal place is a matter of just one run more or less scored per hundred completed innings.

8   Later, he was invested as Sir Vivian Richards by the Queen, a very rare honour that few players have received on their retirement, reflecting the enormous influence that he had on cricket in the Caribbean and the world.

9   "In the Fast Lane". His diary of the 1981 Caribbean tour.

10   It is a measure of Don Bradman's concentration that he was never in 80 Test innings dismissed in the "nervous 90s" and was only once dismissed for a score between 89 and 112: once he got started batting he usually made it count!

11   Averages for anything other than First Class cricket are totally meaningless. In many places you will see averages for One Day matches or, even worse, Twenty20. These numbers are ludicrous. The most worthless averages that I have ever seen are the increasingly frequent tables of averages for T20 series, when said "series" normally consist of just a single match!

One day cricket imposes particular burdens on statistical analysis because the playing conditions are frequently artificial. A batsman will often come in to bat and need to fling the bat at everything, making no effort of conserve his wicket. Remember, to be significant, statistics must be based on comparing like with like. In 40 or 50 over cricket only the top 3 or 4 batsmen in the order will have any chance to build an innings. Any batsman in the middle order or lower middle order will either come in at the end of the innings to swing the bat, or in a deep crisis. You cannot compare the batting average of Luke Wright or Shahid Afridi, who habitually bat at 7 or 8, with that of an opener like Virender Sehwag. In tables of limited overs batting averages the best averages are almost invariably those of the top three in the batting order. In this case the averages are relative and like must be compared with like: the mercurial Pakistani Shahid Afridi averages 23.31 with the bat in ODIs; the England batsman Alistair Cook averages 30.52 – should one conclude that Cook is the better player? No one should not, because Cook has always opened the batting and has time to build an innings, while Shahid Afridi has frequently batted in the middle order or in the tail, with no chance to play himself in before hitting out. One can only compare Afridi's figures with those of other batsmen who bat in the same position in the order as him.

The same applies to bowling averages: an opening bowler attacks for wickets, the second or third-change bowler is primarily interested in containing the batsman and restricting runs, without fielders in attacking positions. A bowler in the middle of an innings will have few close fielders and will tend, of necessity, to bowl defensively to save runs, rather than being allowed to attack for wickets.

[12] At the time of writing, in mid-2009. By late 2010 he had retired, reaching his 800th wicket in his 133rd and last Test match, at a final average of 22.72. It seems certain that no other bowler will ever approach 800 Test wickets. Murali is a phenomenon.

Chapter 4

Ballistics for Beginners
(and batsmen)

Don't bother looking for that, let alone chasing it. It's gone straight into
the confectionery stall and out again.

Richie Benaud on one of Ian Botham's bigger hits
(Headingley, 20th July 1981)

Let's face it; cricket is not immune to its military overtones. Words like "attack", "bombardment", "explosive" and "surrender" litter cricket reporting. Ballistics is the science of trajectories. It does not matter whether the projectile in question is a hand grenade, a Michael Holding bouncer (some would say that the two were synonymous anyway), a Virender Sehwag slash shot, or a skier from a befuddled Chris Martin, it will fly through the air and eventually come to land somewhere. The science of ballistics allows us to study how and where they will come to earth and what, if anything, we can do about it.

Some incidents in cricketing bombardment are so famous that they are ingrained in the memory, such as Brian Close, at the age of 45, being recalled against a rampant West Indian attack and receiving such a brutal salvo from Michael Holding (known as "whispering death" because his footfall was silent on the turf as he raced in to bowl) that even the commentators were flinching as blow after blow peppered his chest[1]. Projecting a hard object that weighs between 155.9 and 163 grams[2] at a speed that ranges from 150-160km/h has overtones of shooting at the batsman.

It is not unknown for a baseball pitcher to claim that he fires the ball in with the force of a rifle bullet. Some fielders in cricket have been known for the power of their throw, although not always its accuracy and there is always the luckless batsman 22 yards from the bowler, who must defend himself and his stumps with a small piece of wood, from a bowler firing the ball at him at high speed.

So, is a cricket ball really as powerful a weapon as a bullet in the hands of a great fast bowler?

Let us have a look at some numbers. Of course, there are many kinds of bullets, ranging from airgun pellets up to the most powerful, high-velocity, high-calibre military ammunition. However, a "typical", medium velocity bullet will weigh about 8 grams and have a velocity of approximately 450m/s. So the kinetic energy of a bullet is:

$$E = \tfrac{1}{2}\,mv^2$$

Where:

E = the energy in Joules
m = the mass in kilograms
v = the velocity in m/s

So, for a bullet

$$v = 450\text{m/s}$$
$$m = 0.008\text{kg}$$

$$E = 810\text{J}$$

What then is the kinetic energy of a Michael Holding delivery?

At his fastest Michael Holding was bowling at around 158km/h, so advantage there to the bullet, which is about ten times as fast. However, the ball weighs a lot more than the bullet.

$$v = 158\text{km/s} = 44\text{m/s}$$
$$m = 0.16\text{kg}$$

$$E = 0.5 \times 0.16 \times 44^2 \text{ J}$$
$$= 155\text{J}$$

In other words, a Michael Holding delivery "only" had about one fifth of the kinetic energy of a medium velocity bullet when it left his hand, but still packed a pretty formidable punch, comparable to what would normally be called a low-velocity bullet. It is only the fact that the point of contact is larger than for a bullet, so the pressure on the impact point is lower, that stops the ball going straight through the luckless batsman when it hits him. However, the energy of the ball is still high enough to cause what may sometimes be severe injuries to a batsman, with broken fingers and, on occasions, broken arms, not to mention cracked skulls and fractured cheekbones an operational hazard.

Why does a cricket ball cause so much damage? The obvious answer is "it is hard and going fast" but, let us look at it more scientifically. When a ball hits the batsman, there is a transfer of energy and momentum and this leads to pressure exerted by the impact.

The pressure exerted by an object is the force per unit of area.

$$P = F/A$$

Where the pressure is measured usually in Pascals, or in bar (1 bar is approximately the atmospheric pressure at sea level)

$$1 \text{ Pa} = 1 \text{ N/m}^2 = 10^{-5} \text{ bar}$$

For the bullet, the energy of impact is approximately 800 Joules = 800 Nm, which is distributed over a cross-section of approximately 5mm diameter, although as the bullet's tip is rounded, the initial pressure at contact will fall on a small smaller area and thus be even greater, hence its power of penetration. So

$$P = 800 \ / \ 0.0025^2 \varpi$$
$$= 41 \text{ MPa}$$

In contrast, the cricket ball's energy of impact of approximately 150J is spread over a cross-section that we can estimate as being of approximately 2cm diameter, so

$$P = 150 \ / \ 0.01^2 \varpi$$
$$= 0.5 \text{MPa}$$

Thus the cricket ball exerts only about one hundredth of the pressure of the bullet on impact. Even so, when a ball from Michael Holding or Jeff Thomson hit a batsman, the pressure exerted is still around 5 times atmospheric pressure, or a very unpleasant 5 kg/cm², leading to pressure bruising if the batsman is "lucky" and a broken bone, if not.

The aim of a cricketer's protective clothing is to protect him from serious injury by absorbing the energy of impact and distributing it over a larger area. In particular, this is important in the shins where the ball would impact bone, unprotected by a layer of shock-absorbing muscle. The traditional cricketer's pads had bamboo canes protecting the shin, covered with impact-absorbing cotton fibre, although now foam is the material of choice. The bamboo cane distributed the force of the impact, and hence the pressure, along the length of the leg, while the foam absorbs the energy of impact. With the improvement in protective clothing, particularly the introduction of helmets, serious impact injuries are, fortunately, less common than in the past.

The days when Dennis Lillie and Jeff Thomson, or the West Indian pace quartet quite literally terrorised batsmen are now long gone. More often though, these days, when we talk about a bombardment in cricket, we are thinking of the batsman hammering the ball into some far distant corner of the ground, or even out of it. More often than not the boot is on the other foot. In recent years it is bowlers who have flinched when Shahid Afridi, or Virender Sehwag, or Sanath Jayasuriya[3], or Brendon McCallum have come out to bat. This though is a relatively modern phenomenon.

Back in the 1950s and '60s, any reporter describing the action in a Test match as proceeding "at a snail's pace" ran the risk of being sued by an angry snail for defamation[4]. When the first 40 over, Sunday League started in England in the late 1960s, a score of 160 in 40 overs was regarded as a reasonable effort and a potentially winning score[5]. Now, in a T20 game, 160 is usually regarded as little better than par for 20 overs on a reasonable pitch for batting. Although there were exceptions, big-hitting was not so common before the 1970s as it is now.

Move the clock forward to the late 1970s and early '80s and suddenly a whole generation of players had appeared who would strike fear into bowlers, rather than the other way around. In the 1981 Headingley Test v Australia, England were famously 135-7 in their Second Innings, 92 behind and had already checked-out of their hotel in the morning when Ian Terrence Botham famously greeted the England number 9 batsman, Graham Dilley with the words "let's give it some humpty." Botham scored 149 from 148 balls as England won a match that had seemed lost.

Prudential World Cup - 20th match, Group B
India v Zimbabwe
India won by 31 runs

ODI no. 216 1983 season
Played at Nevill Ground, Tunbridge Wells (neutral venue)
18 June 1983 (60-over match)

India innings (60 overs maximum)

		R	B	4s	6s	SR
SM Gavaskar	lbw b Rawson	0	2	0	0	0.00
K Srikkanth	c Butchart b Curran	0	13	0	0	0.00
M Amarnath	c Houghton b Rawson	5	20	1	0	25.00
SM Patil	c Houghton b Curran	1	10	0	0	10.00
Yashpal Sharma	c Houghton b Rawson	9	28	1	0	32.14
N Kapil Dev	not out	175	138	16	6	126.81
RMH Binny	lbw b Traicos	22	48	2	0	45.83
RJ Shastri	c Pycroft b Fletcher	1	6	0	0	16.66
S Madan Lal	c Houghton b Curran	17	39	1	0	43.58s
SMH Kirmani	not out	24	56	2	0	42.85
Extras	(lb 9, w 3)	12				
Total	(8 wickets; 60 overs)	266	(4.43 runs per over)			

Did not bat BSW Sandhu

Fall of wickets 1-0 (Gavaskar), 2-6 (Srikkanth), 3-6 (Amarnath), 4-9 (Patil), 5-17 (Yashpal Sharma), 6-77 (Binny), 7-78 (Shastri), 8-140 (Madan Lal)

Bowling

	O	M	R	W	Econ
PWE Rawson	12	4	47	3	3.91
KM Curran	12	1	65	3	5.41
IP Butchart	12	2	38	0	3.16
DAG Fletcher	12	2	59	1	4.91
AJ Traicos	12	0	45	1	3.75

Zimbabwe innings (target: 267 runs from 60 overs)

		R	B	4s	6s	SR
RD Brown	run out	35	66	2	0	53.03
GA Paterson	lbw b Binny	23	35	4	0	65.71
JG Heron	run out	3	8	0	0	37.50
AJ Pycroft	c Kirmani b Sandhu	6	15	0	0	40.00
DL Houghton	lbw b Madan Lal	17	35	2	0	48.57
DAG Fletcher	c Kapil Dev b Amarnath	13	23	0	0	56.52
KM Curran	c Shastri b Madan Lal	73	93	8	0	78.49
IP Butchart	b Binny	18	43	1	0	41.86
GE Peckover	c Yashpal Sharma b Madan Lal	14	18	0	0	77.77
PWE Rawson	not out	2	6	0	0	33.33
AJ Traicos	c & b Kapil Dev	3	7	0	0	42.85
Extras	(lb 17, w 7, nb 4)	28				
Total	(all out; 57 overs)	235	(4.12 runs per over)			

Fall of wickets 1-44 (Paterson), 2-48 (Heron), 3-61 (Pycroft), 4-86 (Brown), 5-103 (Houghton), 6-113 (Fletcher), 7-168 (Butchart), 8-189 (Peckover), 9-230 (Curran), 10-235 (Traicos)

Bowling

	O	M	R	W	Econ
N Kapil Dev	11	1	32	1	2.90
BS Sandhu	11	2	44	1	4.00
RMH Binny	11	2	45	2	4.09
S Madan Lal	11	2	42	3	3.81
M Amarnath	12	1	37	1	3.08
RJ Shastri	1	0	7	0	7.00

The remarkable scorecard from Kapil's match in the 1983 World Cup. Kapil Devi's six sixes were the only ones scored in the whole match and his sixteen fours were more than twice as many as those scored by his ten teammates combined. It can be argued that this innings had major consequences for the whole future of world cricket.

Two years later, India were 17-5 against Zimbabwe in a World Cup match at Tunbridge Wells and seemingly heading for yet another early World Cup exit[6,7]. Undaunted by the continuing crash of wickets at the other end – 77-6, 78-7 – Kapil Dev hammered 16x4 and 6x6 to take India almost single-handedly to 266 and a score that was just sufficient to win the match against spirited opponents who fought all the way. Incredibly, despite Kapil's amazing innings, Zimbabwe only just fell short in the end. As a comparison to Kapil Dev's heroics, twenty other batsmen that day summed just 24 boundaries and no sixes between them.

My own personal favourites though were two players, one of whom I never saw playing a Test and the other who received less exposure than he deserved. Lance Cairns, the New Zealand bowler[8], often batted as low as 10 in Tests so long was the New Zealand batting, but was capable of the most amazing bouts of violent hitting at the end of the innings. I always watched him bat against England with a mixture of excitement and the hope that he would not continue to flail the bowling to all parts for too long[9]. The most exciting hitter of all though and the great hero of my youth was Mike Procter[10], whose batting feats while playing for Gloucestershire were legendary, including scoring a record-equalling six consecutive First Class centuries (although this was obtained during the off-season, when he returned to South Africa to play in their domestic game) and a destructive pursuit of the Walter Lawrence Trophy for the fastest century of the 1979 English season[11].

There is a fundamental difference between scoring a boundary four and scoring a six. To score a six the ball must, necessarily, go in the air and can be caught, thus there is an inevitable risk to the shot. In contrast, a shot driven all along the ground for four is safe, provided, of course, that it has been correctly executed!

Scoring a six is a simple exercise in ballistics. The ball must be hit at a sufficient velocity and with a sufficient angle of elevation to clear the boundary: hit the ball too slowly and it will not fly sufficient distance through the air; give the ball too much elevation and it will stay in the air for a long time, but not travel horizontally and will be cheerfully pouched by most fielders.

If we ignore air resistance, a ball will travel through the air in a parabola, with gravity pulling the ball back to earth inexorably.

If the ball is hit at a velocity "v" and an angle of elevation θ, the hang time – the time that the ball remains in the air – will be given by:

$$t = -\frac{2v}{g} \text{Sin}\theta$$

Where "g" is the Earth's gravitational acceleration of 9.81m/s^2, decelerating the ball in the vertical axis.

The distance travelled is given by:

$$s = ut + \tfrac{1}{2}at^2$$

Where "u" is the initial velocity, "a" is the acceleration and "t" the time.

In the absence of air resistance (see below) the horizontal velocity would constant so, as a first approximation, the distance travelled before pitching is simply:

$$s = ut$$

With "t" the hang time and "u" the horizontal component of velocity:

$$u = v \text{Cos } \theta$$

To maximise hang time and horizontal travel simultaneously, we will see below that we require that

$$\text{Cos } \theta = \text{Sin } \theta$$

Thus, the optimum angle of elevation for hitting a six is

$$\theta = 45°$$

So, the batsman who understands his ballistics, will attempt to launch the ball at an elevation as close to 45° as possible. By doing this, he maximises both his hang time and his horizontal traverse.

When the batsman gets his timing right, the results can be spectacular. In the 2009 T20 World Cup in England, Kevin Pietersen hit the sixth ball of Yasir Arafat's fourth over in the England v Pakistan Group Match a distance of 104 metres. Assuming that he hit the ball at the optimum 45 degree angle, at what speed did the ball leave the bat?

The hang time of the ball in the air is:

$$t = 2v \sin \theta / g$$

While the maximum height reached is

$$h = v^2 \sin^2 \theta / 2g$$

And the horizontal distance travelled is:

$$s = v \cos \theta / t$$

$$= 2 v^2 \cos \theta \sin \theta / g$$

$$= 2 v^2 \sin^2 \theta / g$$

The maximum range is obtained when the differential of the range with respect to angle of elevation is zero, i.e.

$$ds/d\theta = 2 v^2/g \cos {}^2\theta = 0$$

At angles of elevation up to this optimum value you get *increasing* range each time you increase the angle of elevation still further; at angles beyond this optimum value you get *decreasing* range with an increase in angle of elevation. So, effectively, $ds/d\theta = 0$ defines the point at which the sign changes from a *positive* to a *negative* change in range with increasing elevation.

$$\cos \theta = 0 \text{ when } \theta = 90°$$

Hence, we see that the maximum range is, indeed, obtained for $\theta = 90°/2 = 45°$ and that for the maximum range:

$$s = v^2/g$$

Thus

$$v = \sqrt{(s\ g)}$$

If

$$s = 104m \text{ and } g = 9.81 \text{ m sec}^{-2}$$

$$v = 31.9 \text{ m/s}$$

Or

$$v = 115 \text{ km/h}$$

So how high would the shot have gone as it travelled over the outfield? Remember that the maximum height is defined by:

$$h = v^2 \, Sin^2 \, \theta / 2g$$

Given the velocity that we have calculated of 31.9 m/s, an angle of elevation $\theta = 45°$ and g = 9.81 m sec^{-2}

$$h = 31.9^2 \cdot Sin^2 \, 45°/2 \cdot 9.81$$

$$= 25.9 \text{ m}$$

Of course, this is the theoretical result because it assumes that the trajectory of the ball is an exact parabola, with no air resistance, or wind. Obviously, the wind is not a parameter that we can control as a gust may appear at any moment and the wind will almost never stay constant but, what of air resistance? Will it cause the ball to drop unexpectedly short of the boundary?

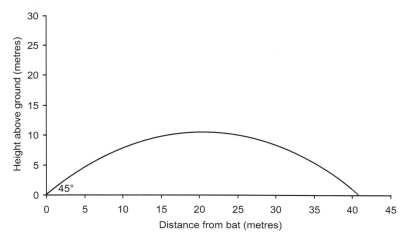

A side-on view of the theoretical trajectory of a lofted shot that comes of the bat at 20m/s (72km/h), lofted at an angle of 45°. The ball will reach a maximum height of 10.2 metres and will land 2.88s after being hit, at a distance of 40.7 metres from the bat. To clear the boundary, the ball has to be hit a lot harder than this! Note how, even hit at an elevation of 45° the trajectory ends up being quite flat, with the ball travelling exactly four times as far horizontally as it does vertically.

Until the ICC decides to arrange cricket matches on the Moon – presumably a future lunar colony will receive first Affiliate and then Associate status and start to play One Day Internationals before playing Tests – leading to changes in match regulations, there will be no cricket played in airless conditions, so air resistance will continue to be a factor on the field of play.

When lunar cricket finally does become a reality, there will be two major influences on play:

1. With lunar gravity one-sixth of that on Earth, boundaries will need to be extended unless scoring is to be limited to being only in sixes. Even mishit shots could fly several hundred metres and big hits may carry 500 metres or more. Spectators will be sat in stands six times as far away from the middle, making following the action extremely difficult without high-power binoculars or large screens carrying pictures of the action taken through big zoom lenses.

2. There will be no air resistance and no wind, so the concept of the high, swirling catch will be eliminated. It also means that the ball will fly in a perfect, parabolic trajectory, rather than being slowed by the wind and air as it flies.

How important is this air resistance to the flight of the ball when there is no wind? Should the batsman calculate the viscosity of the air carefully before considering taking the aerial route to the boundary?

If we are talking about an object flying at low velocity and with a smooth air flow, we can consider a relatively simple case of what is known as Stokes drag, derived by the British physicist Sir George Stokes in 1845.

The drag force is simply defined as:

$$F_D = -bv$$

Where

v is the velocity and b is a constant.

Stokes derived the value for b to be:

$$b = 6\pi\eta r$$

Where, r is the radius of the object and η the viscosity of the fluid, in this case, air.

So

$$F_D = -6\pi\eta rv$$

If this were the case, air drag would not be a problem. The viscosity of air increases as the temperature rises. Between 200K and 500K the viscosity of air doubles. On a typical, warm summer day, with a temperature of 27°C, the dynamic viscosity of air

$$\eta = 1.983 \times 10^{-5} \text{ kg/ms}$$

So, for a standard cricket ball, measuring 7.2cm in diameter, hit at 32m/s, the drag on the ball would be

$$F_D = -6 \cdot 3.142 \cdot 1.983 \times 10^{-5} \cdot 0.036 \cdot 32$$

$$= -4.3 \times 10^{-4} \text{ Newtons}$$

As the force of gravity on the ball is approximately 1.6 Newtons, we can see that the air drag would be negligible.

Simple observation though is sufficient to show that we are not in the happy case of Stokes Drag because, in this case, we would not have the spectacle of a fielder staggering around like a drunk under a high, swirling catch that finally wanders away from him and falls safely to ground. In fact, as such events demonstrate graphically, drag is quite important for a cricket ball.

In fact, we are in a very different regime of drag termed Rayleigh Drag, in which we have a quadratic velocity dependence. In Rayleigh Drag

$$F_D = \tfrac{1}{2}\rho v^2 C_D A$$

Where

F_D is the drag force
ρ is the density of the fluid
V is the velocity of the ball in the fluid
C_D is the drag coefficient
A is the cross-sectional area of the ball

Whether we are in the Stokes or the Rayleigh regime depends on the Reynolds number

$$R_e = \rho v L / \mu$$

Where
    ρ is the density of the fluid
    V is the velocity in the fluid
    L is the characteristic length
    μ is the dynamic viscosity

If the Reynolds number is small we get smooth flow and the Stokes Drag is appropriate. If it is large, we have turbulent flow and we are in a situation of Rayleigh Drag.

What are typical Reynold's numbers?

- For a bacterium, 10^{-5}
- For a spermatozoa, 10^{-4}
- For a tiny fish, 1
- For a delivery from a fast bowler, 2×10^5
- For a swimmer, 4×10^6

In other words, there are ten orders of magnitude difference between the really low-drag case of a bacterium and that of a cricket ball. As we will see elsewhere, the fact that there is turbulent drag on the ball is extremely important to bowlers.

The fact that there is a large drag on the ball means that the ball's terminal velocity is rather low. Suppose that a batsman were to hit an immense slog straight up into the air, how fast would the ball come down to the hands of the luckless fielder underneath?

The answer is 123km/h.

If the slog goes above a certain height in the air it will always fall into the fielder's hands at 123km/h, no matter how far it falls.

The reason is that at this velocity the drag on the ball is the same as the gravitational force. This means that left to its own devices, a falling ball will accelerate for about 5 seconds, falling about 100 metres as it does so, until the drag on the ball exactly balances the force of gravity and will then continue to fall at constant speed and that constant speed is 123km/h. Similarly, as the drag is smaller on a cold day than on a warm day, the ball will fly fractionally further, but batsmen should not look at the frost on the ground and rely on the reduced drag to see their lofted shots clear the inner field, because the effect will be tiny and difference in flight measured in tens of centimeters . If you are listening to yet another England cricketer trying to explain why he has been caught in the outfield rather than see his shot sail over the boundary for six and the player claims that he misjudged the Rayleigh drag on the ball because of the cold, give him full marks for his knowledge of physics, but it is still most likely that the main reason that he was caught in the deep is that he simply mistimed the shot!

Endnotes

[1]   To my amazement, you can even watch some "highlights" (if that is the correct word) of Michael Holding's spell at Brian Close on UTube (http://www.youtube.com/watch?v=w-f5pfBgpNE). Although old, the colour footage is of quite good quality.

[2]   http://www.lords.org/laws-and-spirit/laws-of-cricket/laws/law-5-the-ball,31,AR.html

[3]   None more so than the England bowlers who saw Sri Lanka score 324 to win in just 37.3 overs in 2006 (http://content.cricinfo.com/engvsl/engine/match/225254.html), with Kabir Ali bowling 6 overs for 72 runs, Tim Bresnan 2 overs for 29, Liam Plunkett 5 overs for 46 and Steve Harmison appearing almost economical , at least in comparison, with 10 overs for 97.

[4]   No less than eight times, fewer than 120 runs have been scored in a full day's play in a Test match. Of those, five occurred between 1956 and 1959, four of those at Karachi. The absolute nadir of fast scoring was reached in the 1956/7 Karachi Test between Pakistan and Australia, in which, on the first day, in five and a half hours of play, Australia were all out for 80, with Pakistan scoring 15-2 by the close. Then, on the fourth day, Australia resumed their second innings at 138-6, eventually ending all out 187, to which Pakistan responded with 63-1 in pursuit of 69 to win. The match produced a total of 535 runs for the fall of 31 wickets, in 303.1 overs, at a magnificent 1.76 runs per over.

[5]   In the 1983 World Cup in England, albeit over 60 overs, rather than 40, just three sides averaged a scoring rate above 4 an over.

[6]   Had India *lost* that match, their final group match against Australia would have become the qualification decider rather than a dead rubber and India, the eventual champions, could well have ended up being knocked-out.

[7]   Cricket fans can speculate on how this game and thus this match-winning innings may well have changed cricket history. Before India's unexpected triumph in the 1983 World Cup, the reaction of India's players, administrators and public towards One Day cricket was one of ambivalence at best and total indifference at worst. Had Zimbabwe beaten India that day at Tunbridge Wells and India failed to reach the semi-finals, the sudden explosion of One Day International cricket in India would have been at best delayed by some years and, at worst, might never have happened. The impact on the economics of the game would have been immense and, with it, the political influence that goes with it.

8   In 1983 Lance Cairns had a remarkable summer. In the 1st Test at the Oval, his bowling was pretty unsuccessful, with just one wicket as New Zealand slipped to a heavy defeat, but before they did so, he hit a typical 32 in 20 balls, with 4x6 and 1x4. The 2nd Test at Headingley was, though, Cairns' match: 10 wickets in the match (7-74 and 3-70) as New Zealand won for the first time in England, and won handsomely; there was another, shorter innings too, this time just 1x4 and 2x6, as he scored 24 in 21 balls. Even if he was less successful in the rest of the series, at Lords (where I watched the second day's play) and at Nottingham, that 2nd Test was Cairns' match.

9   A typical Lance Cairns innings was brief and violent. He only scored two fifties in 65 Test innings, with a top score of 64, but his career total of 28 sixes (18% of his Test runs) gives a fair idea of his intentions and entertainment value. In contrast, Geoff Boycott scored just 8 sixes in 193 Test innings (0.6% of his Test runs), almost all from hook shots and another legendary opener of my youth, Sunil Gavaskar, scored only 26 sixes in no less than 214 Test innings (1.5% of his Test runs). Lance Cairns' sixes comprised almost a fifth of all the runs that he scored in Test cricket, which must be one of the highest fractions in Test history.

10   Many people think that Mike Procter, who was a genuinely fast bowler, capable off-spinner, prolific middle-order batsman and superb catcher, would have become the greatest all-rounder in the history of the game had his government been less stupidly narrow-minded in its politics.

11   In "Another Day, Another Match" Mike Procter's teammate and new ball partner, Brian Brain comments how Mike Procter set out in 1979 to capture the Walter Lawrence Trophy and hammered bowling into submission all around the country until he achieved finally the fastest century of the season.

12   A cricket ball is solid and thus has relatively high mass and density, so drag is not so important as it is for games played with a light, hollow ball. In table tennis, air drag is extremely important to how the game is played, as the ball is hollow, thin-skinned and weighs just 1 gram. Anyone who has played table tennis at high altitude, as I used to in the observatory in La Palma, will discover that initially the ball will always fly off the end of the table because the drag is so much lower than at sea level. In contrast, when you return to sea level, it becomes almost impossible to get the ball over the net until you become used to the increased drag! In tennis, with a hollow, but although rather heavier ball, the effect of playing at altitude is also significant and players used to hitting lines with precision at sea level, discover that their game can be quite seriously thrown off when playing at altitude.

CHAPTER 5

NOT QUITE A GOOD REFLECTION

"It just exploded"

John Emburey, Edgbaston 1981

Sunday, August 2nd 1981. Fourth Test. Fourth Day. Australia are 105-4 and, fresh from the remarkable result at Headingley twelve days earlier, in the previous match, where they had failed to chase 130 to win, having had England 92 behind with only 3 wickets left, following-on the previous day, were grinding slowly towards their victory target of 151. Allan Border has been at the crease for over three and a half hours for 40 runs, scored at a strike rate of under 23. In partnership with the debutant, Martin Kent, Border looks set to win the match for Australia. Just 46 are needed to win, with 6 wickets left. England captain, Mike Brearley is becoming desperate: the pitch is easy, the bowlers are not threatening and Allan Border's defences look impregnable. Ian Botham, who has had a very quiet match, taking just one wicket so far and scoring 26 and 3 with the bat, is reluctant to bowl, saying that he does not know how he would take a wicket on such an even-tempered surface[1]. Mike Brearley signals to Peter Willey who, as second spinner, has not bowled yet in this match, to warm-up for the next over. Maybe, just maybe, with the pitch taking some turn, his gentle off-spin could present a new problem to the batsmen but, first, John Emburey has to complete his over. Suddenly a delivery hits a fragile spot in the pitch and produces a large puff of dust like a small explosion and climbs violently towards the batsman's face. Allan Border tries to fend the ball off, but is hit on the gloves and the ball loops to Mike Gatting, crouching at Short Leg, who pouches it safely.

As the England side celebrate, John Emburey wanders around, as if in a daze, repeating again and again "Amazing! It just exploded!" Of course, with Border gone, Ian Botham grabbed the ball and then took the last five Australian wickets for one run in twenty-eight deliveries, finishing with the remarkable figures of 14-9-11-5[2]. Why was that John Emburey delivery so unexpected?

In theory, the trajectory of the ball, on pitching on a hard surface, follows the simple optical rule of angle of incidence equals angle of reflection. Let's make a few assumptions and calculate the ball's approximate trajectory.

The ball was delivered from a height of about 2.5 metres (for a spin bowler who does not have a high, leaping action like a quick bowler, it may even have been a little less), from the crease, one yard in front of the stumps at the bowler's end. The batsman was stood on the batting crease, one yard in front of the stumps that he was defending and the ball bounced on a good length which, for a spinner, is 4 metres in front of the stumps – in other words, about 3 metres in front of the batsman. Let's assume that the ball was speared-in, without any attempt to put flight on it, so that its trajectory was a good approximation to being a straight line from hand to pitch.

A 22 yard pitch coverts to being 20.1 metres long[3], so the ball will pitch approximately 15.3 metres after leaving the bowler's hand. It will be delivered at an angle θ out of the horizontal which is given by

$$\theta = \arctan (2.5/15.3) = 9°$$

The ball will then have travelled 3 metres to the batsman, rising at the same angle of 9° so, the height that it would reach the batsman would simply be

$$h_{bat} = 2.5 \cdot (3/15.3)$$

So, it would have been expected to reach a *maximum* height of 0.5 metres by the time that it reached the batsman, or about two-thirds of the height of the stumps.

The height would actually have been less than this because the coefficient of restitution of a cricket ball is less than one. It would have lost speed on pitching and thus bounced to a lower height than in the ideal case, as the ball will be moving slowly enough after bouncing that its trajectory will no longer approximate a straight line and will actually be a parabola that falls below a straight line trajectory. In any circumstances, the 1.75 metre tall Allan Border should not have had any great problem getting on top of the ball and playing it down safely.

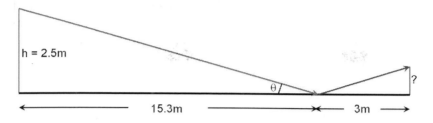

The expected trajectory of John Emburey's delivery to Allan Border if its trajectory had followed, the laws of optical reflection, assuming that the ball was delivered from a height of 2.5 metres and pitched on a good length, 4 metres in front of the stumps. The ball should have bounced to no more than 0.5 metres height, but actually bounced about three times as high.

Flash forward to 1986. The West Indies are at the height of their powers and playing host to an England side that has just beaten India 2-1 in India and then Australia 3-1 in England, winning five and drawing four of the eleven Tests in the two series. There is a strong feeling in English cricket that this side may just surprise the West Indians who, in England in 1984, had blackwashed them 5-0. England went into the first ODI with several players in good form, having just thumped Jamaica in a 3-day match. The West Indies though had to choose which bowlers to leave out from a list of Joel Garner, Malcolm Marshall, Michael Holding, Courtney Walsh, Patrick Patterson and the two spinners, Roger Harper and Viv Richards. After a dreadful start, when Tim Robinson and captain, David Gower were dismissed for ducks, leaving England 10-2, Graeme Gooch and Mike Gatting were standing firm. At 47-2, it seemed that the worst was over. Then Malcolm Marshall pitched the ball short. Mike Gatting went to hook and top-edged the ball straight into his own face and onto the stumps. When the ball was recovered, the fielder made the rather grisly discovery that a half centimetre piece of bone from Gatting's nose was embedded in it. A shaken England lost the match and their best batsman until almost the end of the tour[4].

When the First Test started at the same venue, three days later, they were greeted by the first of a series of hard, uneven surfaces that preyed on their shattered confidence. So devastating were Marshall, Garner and Patterson in that match that Michael Holding – generally reckoned to be one of the greatest fast bowlers in history[5] – was barely required to bowl. Only Graham Gooch, in the first innings, and the ever-courageous Peter Willey – the top scorer on either side in the whole match – in the second innings, passed fifty as England averted an innings defeat by the narrowest of margins.

Faced with four bowlers who were able to bowl with devastating accuracy at 90-95 mph, it is doubtful that England would have been able to put up much of a fight even on perfect, true surfaces. With the batsman uncertain to what height any delivery would bounce, batting became a lottery and the batsmen were psychologically defeated even before coming out to bat. Why though did this happen? What is the difference between a "true" surface and one described as being "like corrugated cardboard"? Why should a ball from John Emburey suddenly leap at the batsman's throat?

The rule of angle of incidence equals angle of reflection still applies. The difference is that we are dealing with a surface that it not flat, like a mirror, but has imperfections so, according to the exact point of impact the incident beam is dispersed on reflection. Consider a case where there is a small ridge in the pitch surface on a good length for a fast bowler – i.e. about 8 metres from the stumps. Let's see what happens to three deliveries that pitch around the position of the ridge.

As the fast bowler's length is much shorter than the spinner's length, the angle of deflection from the horizontal of the delivery is greater. Whereas for a spinner, with a much fuller length, it is around 9°, for a fast bowler it is approximately 12-15° if the bowler is aiming in the range from a good length to somewhat short of a length, so the ball hits the pitch at a steeper angle to start with.

- Delivery 1 pitches a few centimetres short of the ridge. It hits the flat part of the pitch and is reflected at 15° to the horizontal, climbing to approximately waist height.

- Delivery 2 pitches only 10-15 centimetres further up the pitch – a distance small enough that, to the batsman's eyes, the ball has pitched on the same spot as the previous delivery. This though is enough for it to hit the upslope of the ridge, increasing the effective angle of incidence

$$\theta_{reflected} = \theta_{incidence} + \theta_{ridge}$$

Where

θ_{ridge} is the angle of slope of the ridge

The ball will thus be reflected to the batsman at a steeper angle and rise much more, possibly chest or throat high, leading the batsman to protect himself instinctively and potentially to lob an uncontrolled defensive shot into the waiting hands of a close catcher, unless the batsman is quick enough and skilled enough to get the bat above the ball and to angle the blade of the bat downwards.

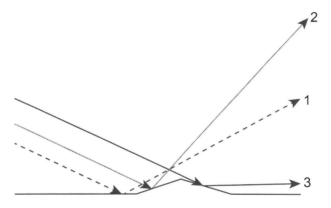

Three deliveries from a fast bowler that land around a good length, on a pitch with a small ridge (here, obviously, exaggerated for effect). Delivery 1 bounces normally. Delivery 2 hits the upslope and bounces much more steeply. Delivery 3 hits the downslope of the ridge and keeps very low.

- Delivery 3 pitches a further 10-15 centimetres up the pitch and lands on the downslope of the ridge. Now,

$$\theta_{reflected} = \theta_{incidence} - \theta_{ridge}$$

In this case the ball is reflected at a much smaller angle to the horizontal and may keep very low (in Australian vernacular, it can "scuttle through like a Bondi tram"). The batsman is prepared to keep down a delivery that he expects to receive at least waist high and possibly leaping at his throat, only to see the ball go through at half stump height, under his defences. Sympathetic word from the commentator "there's not a lot that he could have done about that one" and another batsman trudges off.

From the batsman's point of view, the three deliveries have pitched at what is apparently exactly the same point, but have behaved completely differently. This creates a degree of uncertainty in the batsman's mind that the bowler can exploit: an indecisive batsman is likely to play a hesitant shot and give his wicket away. If, in addition, his confidence is low, due to poor form or poor results, or a poor atmosphere in the dressing-room, it is easy for the batsman to get into a defeatist frame of mind and see his decision-taking adversely affected. Failure breeds failure. If results have been poor, the collective uncertainty can spread through a side like a cancer, which is what happened to England in the Caribbean in 1986.

What about John Emburey's famous explosive ball? The result was similar to scenario 2, but this really was the opposite situation. When it impacted, rather than hitting a ridge, it dug a small crater in a brittle part of the pitch, re-bounding off the firm surface of the upslope of the back wall of the crater at a steep angle, hence it climbed much higher than any such ball had any right to do. When we detect an object coming towards our face at an unhealthy speed we react automatically to protect ourselves: physiologically the batsman cannot stop himself from trying to intercept the perceived threat with the bat. The bowler relies on this automatic instinct for self-preservation to induce the batsman to play a false, or uncontrolled shot.

There is a further factor to consider. Imagine a bowler delivering the ball at 160km/h from a distance of 20 yards – 18.3 metres – away from the batsman. How long would the ball take to reach him?

$$160km/h = 44.4 \text{ m/s}$$

So, the ball would take 18.3/44.4 = 0.41s to reach the batsman.

The human reaction time depends to a large degree on the state of alertness, but is also physiological and dependant on the person concerned. However alert you are, you will not react faster than your own personal speed limit unless you jump the gun. In astronomy this is known as the "personal equation", a term coined by Friedrich Willhelm Bessel in 1823. Bessel noticed that timings by different observers of the passes of stars through telescopes to measure their precise position tended to be systematically different. Some were obviously systematically slower to react than others. This difference could be calibrated quite accurately. In astronomy, where alertness has to be maintained for long periods without a break with the additional inconvenience of cold and discomfort, a personal equation above 0.3 seconds is not unusual and it can be quite significantly higher: for a batsman such a slow reaction time would have the ball almost on top of him before he starts to react. At the other extreme 100 metre sprinters are tensed and ready for at most a few seconds before the starting gun fires and actively trying to anticipate the start. Any reaction time in athletics faster than 0.08 seconds is treated as a false start, but a reaction slower than 0.12 seconds will generally see a sprinter badly left at the start by his or her rivals.

Most people though, in normal situations, will have a hard-wired reaction time in the range from 0.15 to 0.25 seconds. The bad news is that that Michael Holding delivery is already half way down the pitch before you have even reacted to it… and that is before it has seamed or swung after hitting the pitch. In theory, given the speed of release of the ball from the bowler's hand, from pitching to reaching the batsman, the ball takes a lot less than 0.2 seconds to arrive and so no batsman in the world should be able to react in time to any deviation in its path. Why is batting even possible against a fast bowler? The answer is that physics comes to the rescue of the batsman.

We have already seen that the drag force on the ball is

$$F_D = \tfrac{1}{2}\rho v^2 C_D A$$

The drag force depends on the state of the ball. A shiny new ball will have less drag than a bundle of rags 80 overs old because the air will flow more smoothly around it and will generate less turbulence. However, the long and short of it is that the ball will expend 25-30% of its kinetic energy overcoming air resistance and the energy that it loses will be expended setting the surrounding air in motion (i.e. turbulence). If the ball is reasonably new, you can say that it will slow ≈12% by the time that it pitches.

Let's do some numbers, supposing a ball delivered at 160km/h from 2.5 metres height that pitches on a good length, 8 metres from the stumps. We will assume a straight line trajectory, which is actually a pretty good approximation at these speeds. So, we modify John Emburey's trajectory to that of a fast bowler, pitching much closer to the centre of the pitch.

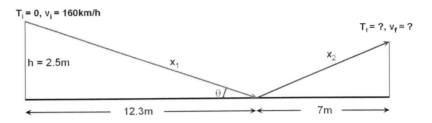

A schematic diagram of a fast bowler's delivery. The ball leaves the bowler's hand at time T_i with a velocity of 160km/h and pitches on a good length, 8 metres from the stumps.

First, how far does the ball travel before pitching? If we approximate the flight to a straight line, the distance travelled is given by the Pythagoras Theorem

$$a^2 = b^2 + c^2$$

Where "a" is the hypotenuse

$$b = 2.5m, c = 12.3m$$

So

$$a = 12.6m$$

The initial velocity of the ball is 160km/h, but air resistance slows it by 12%, so it will have slowed to 141km/h by the time that it pitches, and thus it will cover the 12.6m at an average speed of 150km/h.

The time of flight to pitching is just the distance divided by the velocity, so the ball takes exactly 0.30 seconds to pitch, with a downward deflection of 11°5 (this sounds like a remarkably small angle and it is – only when the bowler drops really short will the angle of deflection from the horizontal get close to 30°).

On bouncing, the ball loses kinetic energy and thus its speed will drop. It will rebound with about 55-60% of its original speed, so the 141km/h reduces to ≈82km/h. Air resistance will slow the ball by a further ≈7% before it reaches the batsman, to 76km/h. So, when the ball finally reaches the bat, it is moving less than half as fast as it started out on release from the bowler's hand. Put another way, the ball is "only" moving as fast a car on a good British "A" road.

The 7m from pitching to the bat are covered at an average 79km/h and take a further 0.32s, so

$$T_f \approx 0.62 \text{ seconds}$$

Even with a delivery at 160km/h, the batsman has about 0.6s to react before the ball reaches him. If it were not for air drag and the relatively low coefficient of restitution of a cricket ball, batting against the very fastest bowlers would be genuinely totally impossible. Even so, the average armchair cricketer, watching from the safety of his position in front of a television at home cannot imagine just how quickly the ball arrives. Put Joe Public in front of Shaun Tait or Shoaib Ahktar and he would probably scarcely even have started to move before the ball crashed into the stumps. Mere mortals who have not played at the highest level can hardly imagine what it is like to face genuinely fast bowling[6]. If we go back to our road scenario above, the batsman has a task similar to a pedestrian who finds himself crossing the road when a he suddenly realises that there is a speeding car only about 25 metres away: you have little time to think and need to react automatically to avoid disaster.

Unlike a member of the public, a skilled batsman will anticipate what the ball is going to do instinctively and will give himself extra time by taking action often before the ball is even delivered. If the main movement is complete early, the batsman has extra time to make the fine adjustment that is needed to accommodate movement off the pitch. If the batsman expects the ball to bounce relatively high, he will take a half step backwards, towards the stumps, to give extra time until the ball reaches him to judge the bounce and to get over the top of the ball. In this case, his weight is on the back foot and the bottom of the bat is at stump height or higher; if the ball fails to bounce as high as he expects it is almost physically impossible for the batsman to react fast enough to transfer the weight back onto the front foot and thus bring the bat down on the ball in time to avoid dismissal. Similarly, if the ball is not bouncing high, the batsman will get forward, killing the bounce of the ball, but at the cost of losing a split-second of reaction time by moving forwards to meet the ball. In this situation, his weight is on the front foot and he is in no position to react and adjust if the ball bounces more than he is expecting: the batsman may be hit by the ball, or may play a completely uncontrolled shot that could fly anywhere. Uneven bounce thus severely reduces the batsman's ability to think ahead and give himself extra time to react by anticipating the ball's trajectory, greatly increasing the danger of dismissal.

For a bowler, bowling a good length means pitching the ball at such a point in the pitch that the batsman is uncertain whether to go forward and meet the ball, or play back and wait for the ball's arrival off the pitch. If the ball is pitched too short, it loses pace on pitching, giving the batsman additional time to play his shot, thus largely eliminating the effect of surprise (bowling that is consistently too short is generally regarded by batsmen as a buffet to be feasted on). If the ball is over-pitched, the factor of the bounce is taken out of the equation, as is to a large degree, lateral movement off the pitch, so the batsman may play his shot with virtual impunity.

Using the seam

One feature of the ball much appreciated by faster bowlers is its seam. We will meet the seam itself in the chapter dealing with the ball. Its effect is to add a further element of uncertainty in the ball's trajectory. The seam is a raised ridge around the circumference of the ball that makes the ball not perfectly spherical. When the ball hits the seam the effect is similar to hitting an irregularity in the pitch as, instead of impacting the curved surface of the ball in a perfectly predictable way, the angle of reflection will depend on the local angle of incidence at the point of impact; i.e. according to which side of the seam hits the pitch, the ball will seem to deviate unexpectedly laterally. If the ball pitches on the left hand edge of the seam, it will deviate to the right; if it pitches on the right hand edge, it will deviate left.

A fast bowler though can also use the seam in a more imaginative way. Legendary West Indian fast bowler, Andy Roberts is one of the best-known exponents of an alternative and less widely used technique of using the seam. Sometimes he would deliver the ball across the seam, such that instead of being upright, it was horizontal and pointing at the batsman. When he did this there were two effects. First, the ball would be less inclined to swing through the air, meaning that its direction could be controlled much more exactly. Second, if the ball landed on the edge of the seam, the local angle of incidence would be increased and the ball would rear unexpectedly from a good length. As the ball would only hit the edge of the seam on a small fraction of deliveries, when it happened the extra bounce would be unexpected and difficult for the batsman to anticipate. Such a delivery was an extremely effective and dangerous surprise weapon for him.

Middlesex and England bowler, Steve Finn, firing the ball down. The seam of the ball is perfectly vertical, giving the delivery every chance to hit the seam and deviate on pitching.

The effect of the seam can be even greater when the pitch is cracked. When the ball hits the edge of a crack, it will tend to move in the opposite direction; combine this with the effect of hitting the seam and the ball may deviate unexpectedly a considerable distance. Even if it appears from the batsman's point of view that the ball is possessed by the devil, it is still obeying the laws of optics in that the angle of incidence is equal to the angle of reflection when we look on a smaller scale. For the batsman though, such effects are totally random and only serve to increase his uncertainty. A highly accurate fast bowler who can pitch the ball consistently in the danger area can make life almost impossible for the batsman, with each ball behaving differently off the pitch.

Hard or soft?

When we say that a pitch is hard and pacey, what does that mean physically? The industry standard for a new ball is that when it is dropped onto a thick steel plate from a height of 2 metres, it should bounce to a height of between 22 and 30 inches (0.56-0.76 metres). Law 8.2 states that

Size of stumps
The tops of the stumps shall be 28 in/71.1cm above the playing surface and shall be dome shaped except for the bail grooves.

In other words, the ball should bounce between 78% and 107% of the height of the stumps when dropped in these circumstances. However, a pitch is not a steel plate. When the ball hits a soft pitch, the soil compresses, individual grains of soil move against their neighbours, which themselves, move against their neighbours and so on. This process absorbs and disperses energy in the same way that a sand bag stops a bullet. The result is that the ball loses far more energy on impact than its coefficient of restitution would predict and comes off the pitch only slowly, giving the batsman far more time to adjust and play his shot than on a harder surface.

On a soft, lifeless surface, the ball loses so much speed after bouncing that, save as an occasional surprise weapon, short-pitched bowling is an exercise in futility, especially with the old ball. A genuinely fast bowler can though create some difficulties by bowling fast through the air and keeping the ball up to the batsman, taking advantage of the fact that the ball will only slow by about 12% before pitching whereas, after pitching, it is likely to slow by as much as 50%. Provided that the bowler does not over-pitch and thus provide easy pickings for the batsman, a genuinely fast bowler can still provide a searching test if he can maintain his pace and accuracy. One of the most famous examples of this was the Oval Test in 1976 when, on a truly lifeless surface, Michael Holding produced the quite stunning match figures of 53.4-15-149-14, while 16 other bowlers in the match had combined match figures of 369.4-74-1253-14[7]. Holding's fiery pace and accuracy allowed him to create difficulties for the batsmen on a surface that left other bowlers impotent.

In contrast, if the pitch is very hard, it will resist compression when the ball impacts it and little of the ball's kinetic energy will be dissipated. In this case the pitch will act in a similar manner to the steel plate used to test the bounce of the ball in manufacture. The main source of dissipation of energy will come from the mechanical deformation of the ball when it hits the pitch. With a hard new ball hitting a hard pitch, the loss of kinetic energy on impact will be minimised, giving a much higher velocity of rebound and hence higher bounce. The batsman will be faced with the twin problems of a ball that arrives much more quickly from release and at higher speed than on a softer pitch and one that bounces consistently much higher.

ENDNOTES

1   A bewildered Mike Brearley, who has just won his second consecutive Test match from a seemingly hopeless position, was asked about Ian Botham in the post-match interview and, still visibly pumped with a huge dose of adrenaline, said excitedly "he didn't want to bowl, you know?". It has been an unusually subdued Ian Botham who had suggested turning to Peter Willey's bowling instead of his own. Ian Botham, in turn, when asked about the situation replied that until the Emburey delivery had exposed a new batsman, he did not know how he would have taken a wicket in the situation as it was. Once Emburey took Border's wicket though Ian Botham knew that suddenly England had a chance and no one would have stopped him from bowling then.

2   Peter Willey never did get a bowl in that match and was then dropped for the last two Tests of the series. In fact, he only bowled 16 overs in total in the four Tests that he did play in the 1981 Ashes series. Fearless hero of the battles against the West Indies in 1980 and 1980/81 (he played 15 of his 26 Tests against them and both his Test centuries were scored against the West Indies, both times to save a match), it would be four years before he played another Test match, returning in time for the disastrous 1986 Caribbean tour. Peter Willey later became a respected Test umpire, standing in 25 Tests and 34 ODIs between 1995 and 2003, before the pressures of constant travel and time away from his family persuaded him to retire from umpiring at the early age of 53.

3   Cricket uses an uneasy mix of old imperial and metric units. The length of the pitch and the size of the stumps have been defined traditionally in imperial units, thus converted into metric the numbers do not come out as neat, round figures.

4   Mike Gatting was sent back to England for an operation to re-construct his nose, returning finally for the 5th Test, in which he scored 15 and 1 as England slipped to a 10 wicket defeat and a 5-0 series loss.

5   He was selected ahead of even legends such as Wes Hall and Joel Garner in the CricInfo West Indies all-time XI (http://www.cricinfo.com/magazine/content/story/468845.html).

6   The difference between village cricket and Test cricket is such that a slow bowler such as Derek Underwood would be faster than most opening bowlers in village cricket. A spin bowler like Shahid Afridi who has a stock ball in the low 70s mph, would be almost unplayably quick in village cricket.

[7] No other bowler in the match took more than 3 wickets despite the fact that the match featured the fearsome pace trio of Vanburn Holder, Andy Roberts and Wayne Daniel on the West Indian side and renowned bowlers such as Bob Willis, Tony Greig, Derek Underwood and Geoff Miller (the current England Chairman of Selectors) were playing for England. The match, won by the West Indies by 215 runs, produced double centuries for Viv Richards and Dennis Amiss and eight fifties in all, although the luckless Chris Balderstone managed a pair, clean-bowled by Michael Holding in both innings, in what was the second and final Test of his career.

This series provides a stark contrast with modern tours. Before the 1st Test, the West Indians had no less than six First Class matches. Two more came in the week and a half between the First and Second Tests and four more between the Second and Third Test. Fringe players, players trying to get into form and those coming back from injury had plenty of chances to impress and to get used to conditions. Counties almost invariably played their strongest available side against the tourists, as the best players tried to make a point to selectors and rivals.

Chapter 6

The Ball

"Well, it's like this: I put it where I like and he puts it where he likes."

Victorian era cricketer speaking of bowling to W.G. Grace

(Grace had just deposited the ball into the lap of a pretty young lady
sitting in a carriage outside the ground)

Law 5 (The ball)

1. Weight and size

The ball, when new, shall weigh not less than 5 1/2 ounces/155.9g, nor more than 5 3/4 ounces/163g, and shall measure not less than 8 13/16 in/22.4cm, nor more than 9 in/22.9cm in circumference.

2. Approval and control of balls

(a) All balls to be used in the match, having been approved by the umpires and captains, shall be in the possession of the umpires before the toss and shall remain under their control throughout the match.

As in so many things in the laws of cricket, the law referring to the ball actually defines very little and relies heavily on tradition. The law does not actually specify either what the ball is made of, or even what colour it should be. The laws only specify the size and weight of the ball. The rest of Law 5 deals with practical issues of ball management during matches. National boards – in England and Wales, the England and Wales Cricket Board (ECB) – have local jurisdiction over the type of ball to be used in the different domestic competitions. For example, for the 2010 English season, the November ECB Board meeting agreed a list of balls to be used in different competitions from the different options offered by manufacturers (http://www.ecb.co.uk/ecb/about-ecb/media-releases/changes-to-2010-rules-and-regulations,308321,EN.html). This led to three different types of ball being approved for different levels of competition: the Duke ball for Test matches and County Championship Division 1 games; the Tiflex ball for County Championship Division 2 and First Class university fixtures; and the white Kookaburra ball for all one-day cricket. Each of these balls has different properties and characteristics that influence how the game is played.

By tradition, the ball for all First Class cricket is red and, by a more recent tradition, since the advent of day-night cricket, the ball for all limited overs cricket is white, whether or not the game is actually a day-night match. Other colours have been experimented with, normally at lower levels of the game, in an attempt to get better contrast in poor light, particularly the twilight period of one day matches before the floodlights have become fully effective and to avoid the problem of discolouration that has plagued the white ball. The one important exception to the red ball rule for First Class cricket was in the 2010 MCC v Champion County match, played in Dubai as a day-night game, for which the playing regulations specified that a pink ball would be used as an experiment, having previously been trialled in a womens' One Day International between England and Australia in 2009. It is fair to say that, although the game was a success and a pleasant change from a freezing Lords in April, not all the players were delighted with the new innovation.

At one point it was even suggested that the England v Bangladesh Test series in Spring 2010 could be played as day-night matches using the pink ball, but the ICC was unenthusiastic about the idea and ruled out the possibility of day-night Tests being approved in the near future. It is probably fair to say that the Bangladesh players, unused to the cool of an English spring day, were probably more grateful than anyone not to have to try to play in the cold of an English spring evening.

If you are a bowler, the sensation of having "the new cherry" in your hand when you mark out your run is a very special one[1]. An article of faith between players and commentators though is that the red and the white ball behave differently, with the white ball swinging far more when new than the red ball. This has led to a trend for bowlers to be classified as "red ball" or "white ball" specialists. In fact, the only difference in manufacture between the two balls is the dye used to colour the leather that covers the ball and the varnish that protects that colour, although, as we shall see, this is an important difference.

While England use the Duke ball in Tests and the First Division of the County Championship and the Kookaburra for all limited-overs cricket, Australia, South Africa and the West Indies use the Kookaburra for all matches (there has been some concern expressed that because of this England may struggle with the different ball in Australia in winter 2010/11 due to their unfamiliarity with it and its less bowler-friendly nature[2]). In contrast in India, the SG, or Sanspareils Greenlands ball is king and has been credited in part with the revival of seam bowling as an art in the sub-continent. Surely though, a ball is a ball and the make of ball matters not a jot? In reality though as we will see, those thirty-eight words that form Law 5.1 allow a great deal of room for interpretation. Only the circumference and weight of the ball are defined and those with significant ranges of acceptable values that translate into significant differences in possible internal composition.

Approximating the ball to a perfect sphere – which it is not!! – we can calculate the acceptable range of ball properties from the parameters given in Law 5.1

The relation between the radius of the ball and its circumference is, of course

$$C = 2\pi r$$

Given that the circumference is in the range from 22.4 to 22.9cm, the permissible radius ranges from

$$3.57 \leq r \leq 3.64 \text{cm}$$

The mass of the ball is given by

$$m = 4/3\pi r^3 \rho$$

Where the mean density of the ball is ρ. The highest possible density comes with the smallest radius (3.57cm) and largest mass (163g) and the lowest permissible density with the largest radius (3.64cm) and smallest mass (155.9g). Thus

$$0.772 \leq \rho \leq 0.855 \text{ g cm}^{-3}$$

This allows a quite substantial range of overall composition to be used and still comply with the regulations on size and weight although, as the density of the ball is always less than 1 g cm^{-3} it will always float if hit into water, at least until it becomes waterlogged[3].

A cricket ball is made with a cork or cork/rubber composite core, tightly wrapped in string, which is then encased in further cork layers and wrapped in leather. As cork has a density of 0.34g cm^{-3} and leather ≈1.35g cm^{-3}[4], in theory less than half of the total mass of the ball can be cork if it is to fulfil the density constraints. Agglometrated cork granules – also known as white agglomerated cork – made a cork granules from 1-3 mm in diameter, has an even lower density typically of just 0.125g cm^{-3}. In practice though, the mixture with synthetic rubber to form the aptly named rubbercork and the tight binding of the core compress it and increase its density such that, in practice, close to 90% of the internal radius of the ball is made up of cork or a cork-based composite.

A cut through a ball designed to be used at good club level. The core is a cork-rubber composite known technically as rubbercork. This is compressed inside a tightly bound wool/polyester layer that is wrapped around the core. Separate sheets of cork wrap around the core and are themselves encased in the leather outer cover that is bound together with heavy stitching
(Image: Platypus Products).

If we take a look at the inside of a cricket ball, we can see how one is constructed and also the range of options that are possible depending on the manufacturing process and components used. There is a huge quality range from the cheapest balls for practice and club matches at the lowest level, through to the quality of ball used in international cricket. The highest quality balls are quarter balls, i.e. the cover is made of four pieces of leather stitched together. Half balls though tend to be cheaper to manufacture and are used at lower levels of the game.

Working from the inside out, we have the inner core of the ball. This is usually not a solid piece of cork but, as we have seen it is instead a mass of cork chippings, often mixed with rubber into a composite[5]. This combination, frequent in composites for industrial applications, allows the compressibility of cork to be combined with the high resilience of rubber.

Cork, harvested from the cork oak (Quercus suber L[6]), has a high degree of compressive elasticity: when deformed by compression, such as when it is used as the stopper of a champagne bottle: it is capable of returning to its original shape on extraction and is also highly hydrophobic – it does not absorb water.

The compressive elasticity of a substance is defined by its Bulk Modulus

$$K = -V\, \partial P/\partial V$$

Where P is the pressure in Pascals, V is the volume and $\partial P/\partial V$ is the partial derivative of pressure with respect to volume. The units of Bulk Modulus are Newtons per square metre. The inverse of the Bulk Modulus ($1/K$) is the compressibility of the material.

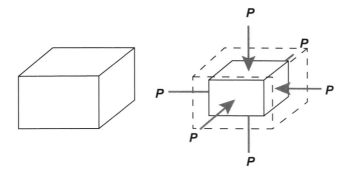

Isostatic pressure deformation for Bulk Modulus determination. Pressure is applied evenly over the surface of the substance to be tested and the decrease in volume is measured as a function of the pressure. The slope of the pressure-volume regression gives the compressibility of the material (adapted from an image design by Christophe Dang Ngoc Chan)

For many substances the compressibility is extremely low. For water it is $2.2 \times 10^9 \text{Nm}^{-2}$. In practical terms it means that at a depth of 4000m in the ocean, where the pressure is $4 \times 10^7 \text{Nm}^{-2}$, the volume of water is compressed by just 1.8%. For steel, at the same pressure, the degree of compression would be just 0.025%. In contrast, agglomerated composition cork has a bulk modulus of $1.6 \times 10^6 \text{Nm}^{-2}$, making it five orders of magnitude more compressible than steel. Forming a composite with rubber actually somewhat reduces the compressibility, to $1.1 \times 10^7 \text{Nm}^{-2}$, but rubbercork remains a highly elastic, strong and lightweight material.

This aggregate is tightly bound inside a layer of spun fibre (wool, string, or similar) that serves to compress it and give it the requisite degree of hardness, without which the ball would not retain its shape. Similarly, without the hardness, as the ball as a whole has only a very low degree of elasticity and thus capability to deform, store energy and then release it to return to its original shape, the shape of the ball would simply distort on pitching. This distortion would absorb the impact energy and thus make the ball lose speed considerably, hence producing only extremely low bounce; it would be not much better than trying to play with a ball made of plasticine. A soft ball also comes off the bat slowly for the same reason – the impact of the bat distorts the ball, absorbing its kinetic energy and converting it into heat – making run-scoring difficult, even if the batsman can cope with ankle-threatening bounce. The harder the ball, the better its bounce will be.

One of the biggest problems that ball manufacturers face though is in ensuring that balls retain their hardness. It is natural that the constant hammering of the ball against the pitch, against the bat and, if hit hard enough, on the concrete of the stands, should soften the ball. Some types of ball soften and thus become more benign faster than others. It is generally reckoned that the Tiflex ball used in the Second Division of the County Championship does tend to get soft more quickly than the Duke or Kookaburra ball used in other English competitions, although it retains its shape well and instances of having to change a ball because it has gone out of shape have become mercifully rare in recent years.

The hard, inner core of the ball typically weighs from 95 to 100 grams. One of the techniques used to ensure that it is as hard as possible and does not lose its shape is to use damp wool or string to wrap the cork. As cork is hydrophobic it does not absorb the humidity but, as the fibre dries, it contracts and holds the core even more tightly inside. However, this also implies that the core must be given time to dry thoroughly before it can be encased in the exterior layers of the ball.

Surrounding the spun fibre layer we have further layers of cork. Unlike the core, this layer is made of cork sheets that fit smoothly around the layer of fibre, giving the outer shell of the ball a solid seat. Finally, sat on top, is the thick leather outer coating.

Obviously, leather is neither naturally red, nor highly polished so, before it can be stitched around the core, it must be dyed. The porosity of leather and its ability to soak up a wide variety of different liquids, both water-based and organic solvents such as oil, allows the coloration to soak completely through, so the colour is retained even if the surface of the ball becomes scuffed and abraded. Traditionally cricket balls have been dyed with derivatives of Rubia tinctoria – the Madder plant, or its Indian variant, the Indian Madder, or Rubia cordifolia. The Madder is a relative of the coffee plant and has been widely used as a source of red dye for at least five thousand years[7]. Although madder root is yellow when first cut, it quickly oxidises to a deep red colour on exposure to air, giving a dye that goes by the generic name of Alizarin crimson[8].

A worn quarter ball showing the internal stitching around the meridian that has ceded with wear and the six lines of stitching around the equator, forming the seam of the ball.

The active ingredient of madder plant dye is, as its name suggests, Alizarin. This is a simple aromatic molecule known as 1,2-dihydroxyanthraquinone, or $C_{14}H_8O_4$, consisting of three carbon rings with oxygen filling the unoccupied carbon bonds on the central ring and two hydroxyls doing the same on one of the two end rings. There are no fewer than ten possible isomers of this molecule, of which Alizarin is just one. It is derived from anthraquinone, $C_{14}H_8O_2$, by substituting hydroxyl for hydrogen as a functional group on adjacent atoms in the side ring. Curiously, anthraquinone itself is not red, but is, in fact, a bright yellow crystalline substance that is only slightly soluble in water but, as a non-polar molecule, is highly soluble in hot organic solvents.

The Common Madder (Rubia Tinctorum). The roots of this plant have traditionally produced the Alizarin dye used to colour the leather of cricket balls.

As alizarin is a rather simple organic molecule and was thus one the first natural dye to be synthesised artificially, artificial alizarin can be produced very simply chemically, largely replacing the need for the natural product. This is done by oxidising anthraquinone-2-sulfonic acid with sodium nitrate – better known as saltpetre – in concentrated sodium hydroxide. Chemically, this reaction can be described as a two stage process:

$$C_{14}H_8O_5S + O \rightarrow C_{14}H_8O_6S \rightarrow C_{14}H_8O_4SO_2$$

With the sulpho group (SO_3H) being replaced by two hydroxyl groups.

Aside from being a dye, it is also used medically to stain tissue samples as it is an effective tracer of calcium phosphate – i.e. bone.

Every cricketers' friend: the alizarin, or 1,2-dihydroxyanthraquinone molecule that gives the red colour to the dye used traditionally to colour cricket balls. The molecule consists of three linked rings of six carbon atoms; the central ring has an oxygen atom on each side replacing the hydrogen atoms around the ring, while the end ring replaces two of the hydrogen atoms with hydroxyl groups.

Alizarin though, fades progressively through photodecomposition. We can see this as the characteristic bright cherry red colour of a new ball darkens with time and use to a much less visible reddish purple. So, for other applications, such as weather resistant paints and coatings where progressive fading on exposure to light is undesirable, alzarin has been replaced by the more complex and photostable quinacridone, a molecule with five carbon rings in a linear formation, with an NH and O functional group on the second and fourth ring respectively.

Not all leather is though suitable for cricket balls. Although top quality leather without scratches or marks is selected, when the leather is cut into sheets and dyed, not all leather will absorb the dye in a consistent manner. Before it can be cut and shaped it must be inspected closely to ensure that the dye has been evenly absorbed and that the colour is without blotches or blemishes. The dyeing process itself may vary considerably in sophistication.

In its simplest form it is a cottage industry with leather being treaded in dye vats by people in rubber boots. At the other end of the scale, enzyme treatment of the leather prior to immersion in the dye has been found to increase considerably the rate and degree of dye take-up, as has the use of ultrasound. Experiments show that when leather is submitted to ultrasound in the dye tank, the leather absorbs a large fraction of the sound energy, leading to acoustic cavitation and a greatly increased absorption of dye. In contrast, agitation on its own of the leather in the dye tank has no effect whatsoever on the rate of dye take-up.

For higher standard cricket, four equal segments of leather are used forming two orthogonal lines around the ball forming an equator and a meridian circling the ball; this is a quarter ball. Once the leather has been dyed, inspected and passed as suitable for use, it is cut into the segments that will be shaped around the core. The segments are like the peel of a quartered orange: roughly oval, with a pointed apex at each end. The four segments are stitched together to encase the cork, but this stitching is done in a radically different way for the meridian to the way that the equator is stitched, as we can see in an old, worn ball. Around the equator the stitching is external, with multiple lines of thread. In the meridian, it is internal and thus not visible from the outside when the ball is new. The two pairs of segments are joined together along the meridian line using a line of stitching that does not penetrate through the outer skin, taking advantage of the considerable thickness of the leather, to form two, smooth hemispheres. In fact, in the highest quality of ball the join should be almost imperceptible and there should be no evidence at all of internal stitching when the ball is new. As the ball wears, the stitching around the meridian tends to cede and, in extreme cases, may even burst, making the ball unusable[9].

Finally, the core is then encased in the two hemispheres of leather, which are pressed against it under intense pressure. Even at this stage the manufacturer can apply considerable variation to the final product. A thin, narrow strip of leather goes around the core under the join between the two hemispheres. This strip gives the ball the slightly raised seam and anchors the six rows of stitching – three on each hemisphere – that are used to seal the ball closed. The laws permit a minimum of 78 and a maximum of 82 stitches to be used around the circumference of the ball, thus the size of the stitches is tightly controlled, but some leeway is permitted, allowing the characteristics of the ball to be varied slightly by the number and size of stitches used. Manufacturers though can still apply considerable variation to this seam by varying its height using a thicker or thinner leather strip underneath it, by using a heavier or a lighter thread for the stitching of the seam (thicker thread also raises the seam, making the ball more responsive to seam bowlers and gives a better grip in the hand). Other methods include treating the thread to make it stronger (for example, in India it is coated with animal fat, increasing its tensile strength, while the change to a thicker thread in SG balls a few years ago has undoubtedly helped Indian seamers to become a far more potent force in home Tests and contributed to the side's fast rise in the Test rankings). Similarly, a different product may be obtained by using mechanical or hand-stitching (in general, hand-stitching is regarded as the superior product, although more expensive as it requires highly skilled craftsmen). Some manufacturers may also use glue to aid the bonding of the leather as well as stitching.

Once the stitching is complete and the ball is sealed, the ball is stamped with the manufacturer's name and brand mark and a lacquer may be applied to protect its finish.

At this point the finished balls go through a final, rigorous quality control inspection. Manufacturers are extremely reluctant to reveal details of their processes, but the one manufacturer that has been willing to reveal quality control information – the Indian manufacturer, SG – freely admits that despite constant intense quality control throughout the entire manufacturing process, 10% or fewer of their finished balls have a sufficiently perfect finish to be regarded as being of Test match quality[10].

Although the laws do not enforce a quality control standard for the hardness or bounce of balls, there are two quality standards that are applied throughout the cricketing world: one used by umpires and the other by manufacturers. The umpire carries a metal hoop through which the ball must pass to verify its size and that it is not out of shape. A ball that will not pass through the hoop must be replaced. For manufacturers, there is a hardness standard that, while not set down in the laws, must be demonstrated to be fulfilled for balls to be accepted for use.

Obviously an umpire would not damage a new ball by throwing it into the pitch and checking that it bounces; pitches are of such variable properties that such a test would be meaningless anyway. Instead, the large Bulk Modulus of steel is put to practical use to provide a consistent reference surface. A sample ball or balls from each batch is/are dropped from a standard height of 2 metres onto a heavy steel plate and the height of bounce is measured.

The basic equations of motion say that

$$s = ut + \tfrac{1}{2} at^2$$

Where "s" is the distance that the ball moves in time "t", "u" is the initial velocity of the ball and "a" the gravitational acceleration.

As the ball is released from rest

$$u = 0$$

So

$$s = \tfrac{1}{2} at^2$$

And hence, the time taken to fall is

$$t = \sqrt{(2s/a)}$$

As s = 2m and a = 9.8ms^{-1}

$$t = 0.64s$$

The velocity of impact

$$v = u + at$$

So

$$v = 9.81 \times 0.64 = 6.26 \text{ ms}^{-1}$$

For a ball to be accepted for use it must bounce to a height between 0.56 and 0.76 metres. Any higher and the ball deemed to have dangerously high bounce, particularly in the hands of a fast bowler on a hard pitch. In contrast, a ball that fails to bounce to the minimum height set may be acceptable when new and hard but, as it wears, it will soften to such a degree that the ball will barely bounce and fluid strokeplay will become well nigh impossible.

Given that such a wide range of properties of the ball are possible and that one new ball may bounce up to 35% higher than another, seemingly identical one on the same pitch, we should not be entirely amazed if, on changing the ball that is being used, the new ball plays differently to the old one, even if both are used balls, especially if they are from different batches. Similarly, if on the first day of a match the ball comes through slower than expected from an inspection of the pitch and bounces lower, the problem may not be a misreading of the pitch, it may be that the ball has a coefficient of restitution that lies towards the lower limit of what is acceptable. Curiously though, it is always the pitch itself that receives the blame by players and commentators alike. We tend to forget that a much wider range of bounce (35% over the minimum permitted) is permitted in cricket balls than is permitted than for a tennis ball (<9% over the minimum).

Cricketers also tend to be more tolerant and restrained souls than tennis players: a cricket ball that bounces less than the bowler expects is treated with resignation; a tennis ball that does not conform to expectations will be roundly condemned by the players and may lead to threats of a boycott[11]. Although as Cassius pointed out[12], anticipating one of Ted Dexter's more memorable faux pas by a mere 2000 years[13]:

> *"The fault, dear Brutus, is not in our stars, but in ourselves that*
> *we are underlings…"*

A suffering opening bowler who is unexpectedly not producing the goods in apparently favourable circumstance might be advised to respond to his captain, paraphrasing Cassius with the words

> *"The fault, dear Andrew, is not in ourselves, but in our balls that*
> *we are wicketless…"[14]*

All in all though, the ball is a remarkable combination of technology, materials science, physics, biology and chemistry.

Endnotes

[1] As a new ball for First Class cricket costs around £70 (about $100), and an absolute minimum of three are required for a two innings match (if one side wins by an innings without ever requiring a new ball to be taken), they are a considerable luxury at the lower levels of the game. For the County Second XI Championship sides are actually permitted to use any ball licensed by the British Standards Institute. A typical hard-up village side may only be able to afford to buy two or three balls per season. In my two summers of playing with Addington Seconds, a new ball only fell into my hands once! "Look after it" said the captain. In fact, I recall that it was the only match that I played in for them in which the game did start with a new ball. It felt like desecration to use it.

In theory, a side could open the bowling with an old ball – for example, for the fourth innings on a worn pitch – but, in practice, either captain can demand a new ball be used at the start of an innings so, in this hypothetical case, it would be the batting captain who would take the new ball to avoid the danger of being spun to defeat quickly on a disintegrating pitch. Bizarrely, on the fourth day of the Sixth Test v Australia at Sydney, on February 14th 1978, the umpires rejected Mike Brearley's request for a new ball to start the innings with England chasing 34 to win on the grounds that they could not find the pertinent law (it is (B) 5 in the Laws of Cricket). This refusal led to leg spinner Jim Higgs sharing the new ball with off spinner Bruce Yardley: a quite singular opening bowling partnership. England finally won by 9 wickets after a considerable struggle.

[2] The reader has a big advantage over me as he or she will know what has happened, whereas I have no idea. Before the 2006/07 Ashes series no one could really have predicted what was going to happen and how the series and the winter would end. The reader will know whether England's bowlers have been made to look toothless and the team, rudderless or, in contrast, they have strode through the series sweeping all before them.

[3] Even though cork resists water, leather and twine soak it up and become waterlogged and denser.

[4] There is a range of values of about ±0.06 g cm^{-3} in the density according to the type and quality of the leather.

[5] Some manufacturers use the less expensive cork sheeting rather than granulated rubbercork. This is one of the reasons why cricket balls differ in characteristics and behaviour from manufacturer to manufacturer and from country to country.

6   The cork oak is principally found in the Mediterranean region. Portugal accounts for 51% of world cork production, followed by Spain (26%), Italy (7%) and Morocco (6%) [1995 figures], with the rest of the world contributing just 10%.

7   Items dyed with madder plant dye were found in the tomb of Tutankhamen.

8   By tradition this dye was used to colour the coats of British soldiers, giving them their traditional name of The Redcoats. There is also some evidence that it was used to dye early American flags of War of Independence vintage, raised in the battles against the Redcoats. There are many varieties of the plant growing in temperate areas of Europe, the Americas and Asia.

9   The laws of cricket contemplate the possibility of the ball splitting in two on being hit due to a failure of the stitching. Were this to happen, the umpires should call Dead Ball and even if part of the ball has flown over the boundary, no runs may be accrued from the shot.

10   There is a wonderful article on how Sanspareils Greenlands manufactures its balls by hand, using traditional techniques, in http://www.cricinfo.com/wac/content/story/210922.html. In contrast, manufacturers in other countries pointedly ignored my requests for information: this is all the more surprising given the amount of freely available material in the scientific literature about cork, cork composites and working leather!

11   Let's face it. The sort of outburst protagonised by cricket's current bad boy, Stuart Broad, which lead to demands for him to be banned forthwith, would not even be rated worthy of mention in a report on an average tennis match.

12   Julius Caesar (I, ii, 140-141).

13   While trying to explain yet another England disaster against an Australian side that the side had been expected to beat, Chairman of Selectors, Ted Dexter, memorably suggested that maybe an errant planetary alignment was to blame for England's misfortune.

14   This is not an entirely whimsical suggestion. The England side has had a tradition of including Oxbridge-educated classical scholars. Vic Marks included various Latin quotes (without translation) and references to Chaucer and to Greek literature in his book on the 1984/85 tour of India and Sri Lanka ("Marks Out of XI" –the best tour account that I have ever read) claiming that he wished to use the knowledge somehow and that he expected his English Cricket Board censor to be capable of checking these references! And Frances Edmonds, former England spinner Phil Edmonds' wife, once reprimanded a heckler in

the crowd in a tour match of the Caribbean in 1986 with a stern comment in Latin! (the incident is recounted in her book "Another Bloody Tour". The heckler was, apparently, not convinced by this intellectual defence of Tim Robinson's batting.

CHAPTER 7

ARMS RACE – THE BATSMAN'S WEAPON

Tempora mutantur et nos mutamur in illis

Times change and we change with the times – the cricket bat's motto

LAW 6 (THE BAT)

1. Width and length
The bat overall shall not be more than 38 inches/96.5cm in length. The blade of the bat shall be made solely of wood and shall not exceed 4 1/4 in/10.8cm at the widest part.

2. Covering the blade
The blade may be covered with material for protection, strengthening or repair. Such material shall not exceed 1/16 in/1.56mm in thickness, and shall not be likely to cause unacceptable damage to the ball[1].

3. Hand or glove to count as part of bat
In these Laws,
(a) reference to the bat shall imply that the bat is held by the batsman.
(b) contact between the ball and either (i) the striker's bat itself
or (ii) the striker's hand holding the bat
or (iii) any part of a glove worn on the striker's hand holding the bat shall be regarded as the ball striking or touching the bat, or being struck by the bat.

© Marylebone Cricket Club 2003

Ball games are essentially of two types. In the first type the players use their own anatomy – hands, feet, head, arms, etc. – to propel the ball into a goal of some kind. Football, rugby, handball, basketball, volleyball, water polo, American football, fives, or even the wizarding game, Quiddich, fall in this class[2]. The goal may be small or large, round or rectangular, up high or defined as part of the surface of the playing area, but the idea is essentially the same in that the player's body applies the force directly to the ball. In the second type of ball game, which includes cricket, to score the players use a bat of some kind to propel the ball with more force and at greater speed and with more accuracy than human anatomy can manage alone. This class of games includes tennis, cricket, hockey (field and ice), baseball, golf, badminton, table tennis, squash and many others and is achieved by applying simple principles of physics and mechanics such as the lever.

Over the years, technology has been increasingly applied to the problem of propelling the ball used in different games with the maximum force and precision. Fifty years ago all tennis players would use a racquet with a wooden frame, strung with cat gut. Now, no serious player would take to court without a racquet of carbon fibre or some similar composite that Fred Perry, or Rod Laver would not even have recognised, although the shape of the racket is essentially unchanged and, from a distance it looks identical to its forebears. The advantages provided by new materials are such that a top player who tried to use a wooden racquet these days would soon see his ranking slump. In contrast, although technology has increasingly invaded cricket, the bat has remained astonishingly fundamentally unchanged for centuries and its composition and characteristic silhouette just as recognisable in 2011 as it would have been in 1811. The reason for this is that, in the laws of the game, the law governing the bat was laid-down in 1774 and has barely changed in nearly 250 years!

The laws of cricket, as enshrined by the Marylebone Cricket Club, actually say very little about the bat, as can be seen in Law 6 (The bat), reproduced above. Only the maximum length, the maximum width and the composition are defined. There are more words about role of the batsman's hand (80, in Section 3) than there are about the bat itself (75, in Sections 1 & 2 combined). In some ways the cricket bat has changed completely, but its composition and basic structure are exactly the same as they were in the 18th Century. In part, these changes in the design of the bat have responded to changes in the bowling technique used and in part due to an increasing use of technology, but the law referring to the bat itself has, surprisingly, changed little. This is mainly because tradition has been respected in a way that it has rarely been respected in other games.

In fact, only one significant change has been made in the laws referring to the bat since 1835. That change was made as a result of a famous incident during the 1979/80 England tour of Australia, when the Australian fast bowler Denis Lillee came out to bat against England with an aluminium bat. The England captain, Mike Brearley, protested that the bat was damaging the ball; whilst the Australian captain, Greg Chappell, protested that Lillee's first scoring shot – a drive down to long off from an Ian Botham delivery – would have been a four if Lillee had used a proper bat, instead of pulling up as it did short of the boundary![3] Kim Hughes brought a wooden bat out, which Lillee refused. Finally, after a long and heated discussion with the umpires and the intervention of Greg Chappell, Lillee flung his aluminium bat into the outfield and reluctantly agreed to continue using a wooden bat. This incident ultimately led to a revision of the laws to stipulate that the blade of the bat should be made of wood.

Before the stipulation that the blade of the bat should be made of wood was added, the previous change to the laws had been made in 1835, when the length of the bat was limited to 38 inches (96.5 cm). The maximum width of the bat was fixed in the 1774 laws, after one enterprising player, apparently Thomas "Daddy" White, playing for Chertsey against Hambledon, decided to use a bat that was actually as wide as the stumps themselves[4].

Notice though that neither the type of wood to be used, nor the shape of the blade, nor the length of the handle, nor the thickness, nor the weight of the bat are anywhere defined[5]. This leaves plenty of scope for innovation so, in reality, it is surprising that the cricket bat has not seen even greater changes than it has. A cricketer from, say, the Hambledon team of 1790, would certainly recognise a cricket bat from 2009, although some aspects of its appearance would seem strange to him, notably the thick edges and scooped back of the blade. Similarly, a current Test batsman would be able to bat with a cricket bat from 1790, although its weight and balance would seem odd[6].

The only fundamental change in the design of the bat came as far back as the mid-18th century to reflect a change in the style of bowling. Prior to this, "bowling" meant precisely that: the ball was rolled, or bowled along the ground to the batsman. With no bounce to worry about, the bat was shaped like a hockey stick and shots were also in the main rolled along the ground rather than being lofted. In fact, there is speculation that the game of cricket and the original shape of the bat itself came about from a game played by shepherds with their crooks. In the 1760s, although bowling continued to be underarm, the ball was, for the first time, pitched or lobbed at the batsman. The distance from the bowler where the ball pitched and it subsequent bounce introduced a new problem for the batsman. With this mode of delivery, the hockey stick bat shape, with its thin horizontal blade became totally impractical and the current, approximately rectangular, vertical blade became necessary, otherwise hitting the ball became almost impossible. Thus, as in any kind of warfare, a major change in the bowler's offensive technique, led to a necessary change in the batsman's defensive armament to counter and neutralise it, giving a new balance of forces at a higher level.

The oldest known cricket bat in existence. Preserved in the museum at The Oval (Kennington, London), the bat dates from 1729. This was the epoch of underarm bowling, when the ball was rolled along the ground. Similar to a hockey stick in shape, such a bat became totally outdated by the change in bowling technique to lobbed deliveries. Note that the wood is much darker than for modern bats; although wood does darken with age this bat was made from heartwood willow, denser and darker than the sapwood willow used today. Image: wikicommons; author, SmokeDog.

The later change from underarm to round arm and on to overarm bowling only changed the degree of bounce obtained with the ball and did not require a change in the bat, only in the batting technique employed, hence the laws referring to the bat have required only minor adjustments since 1774.

Despite the lack of gross changes in design, technology has impinged ever more on the bat. No longer is it a simple, flat prism of wood of almost rectangular base, attached to a handle, with the whole thing weighing close to 1kg. The modern cricketer can choose from a bewildering of designs: long and short handle, traditional and scalloped centre with thick edges, straight or angled shoulder, normal-sized or reduced blade. In general though, the trend is that bats have got progressively heavier over the last 40 years.

In the early 19th Century it was not unusual for batsmen to use a bat as heavy as 5lb (2.25kg). However, by the 1920s, the best batsmen were using a bat weighing just 2lb 2oz (0.95kg): Jack Hobbs, Don Bradman and Wally Hammond all used bats of this weight with great success (as did Peter May in the 1950s), as can be seen from their career figures in Tests.

- Jack Hobbs's average of 56.94 in Tests, with 15 centuries and 2 double centuries over a long career of 61 Tests between 1907 and 1930 marks him as one of the all-time greats.

- Wally Hammond though makes Hobbs look almost pedestrian. Despite missing seven possible years of Test cricket to the Second World War, in 85 Tests between 1927 and 1947 Hammond averaged 58.45, with 22 centuries, 7 double centuries and a triple century[7].

- Don Bradman played fewer Tests than Hobbs or Hammond, but his famous batting average of 99.94, his 29 centuries (in only 52 Tests), 12 double centuries and 2 triple centuries are so far ahead of any other batsman in the history of the game that the difference is positively insulting[8].

There were some exceptions who used a much heavier bat, for example, Australian Bill Ponsford used a bat that weighed 2lb 9oz (1.15kg[9]), which was affectionately christened "Big Bertha". By today's standards though, Bill Ponsford's bat still would be considered lighter than average. It was Clive Lloyd, the legendary figure of the great West Indian teams of the 1970s and '80s, who was one of the pioneers of the larger and heavier bats that are in common usage today. Lloyd's bats exceeded 3lb (1.35kg) in weight. This was topped by Ian Botham who would use bats as heavy as 3.5lb (1.55kg): unsurprisingly, Ian Botham was one of the biggest hitters in the game.

Modern legends use bats of very different characteristics. Sachin Tendulkar uses a 3lb (1.35kg) bat, as did the magnificent Sri Lankan, Aravindra de Silva who, at 5ft 3½in (1.61-m) tall was one of the shortest top-class batsmen in the game. In contrast, Steve Waugh used a much lighter bat at 2lb8oz. So, let us start by asking why bat weights suddenly dropped so much in the late 19th Century before examining why heavier bats have come back into vogue. To find an explanation for the former we have to look at the raw material for bats itself.

Bats are traditionally made of willow (Salix Alba, or the White Willow[10]). A specific sub-species, Salix Alba Caerulea, sometimes called the Cricket Bat Willow, is cultivated in the British Isles to produce the high-quality timber needed in bat manufacture. For bat manufacturers this sub-species of uncertain origin is distinguished by its rapid growth and straight trunk, although to the botanist, the distinguishing factors from other sub-species are the size and colour of the leaves, which are larger and bluer than for other varieties.

The white willow (Salix Alba). Although photographed in the German countryside, this tree is very similar to the English sub-species (Salix Alba Caerulea) specifically cultivated in the British Isles for the timber used in cricket bat manufacture. The biggest difference is that willows grown for bat manufacture are specially cultivated and maintained to ensure that they grow a straight trunk with no branches, except those at the top of the trunk, unlike this wild example. Without continuous maintenance in their initial growth the result would be a tree like this one, totally useless for bat manufacture. (Wikicommons; author, Willow)

To ensure that the cultivated wood is of the highest possible quality for bat manufacture the trees destined for bat wood are specially maintained, especially for their first five years of growth until the bark has hardened. Knots in the cut timber dramatically reduce its quality, so the trunk of the sapling must be carefully and regularly stripped of shoots that would grow into branches; without this continuous maintenance a sapling would grow like the pictured willow, totally unusable. An unmaintained sapling will pass from having a clean trunk to an unusable one in a year or less.

Willow has an interlocked grain that makes it resistant to splitting – most definitely an advantage when it is being subjected to constant impacts! The cells spiral around the trunk, reversing direction every few growth rings. This produces ribbon figure in the grain that can be seen clearly when the wood is cut. Willow wood though has very different properties according to where it is cut from in the trunk. Heartwood, from the centre of the trunk, which was what was used initially to make bats, is both darker – described by furniture makers as reddish brown – and denser (when dry, about 600kg m^{-3},[11]) than the sapwood from around the exterior, just below the bark, which is more creamy white in colour and, when dry, only 390kg m^{-3},[12]. Even a cursory glance at a cut willow trunk will reveal the considerable difference in colour between the dense central third of the diameter approximately and the more spongy outer parts of the trunk.

The number of grain patterns across the blade will be a minimum of four. For the highest grade of bat there will normally be just four straight lines of grain; a batsman will normally specify the number of grains that he requires however, this is purely cosmetic. The number of grains and their straightness, in fact, makes absolutely no difference either to the mechanical properties of the wood or to the way that the bat plays.

As wood is classed as being a hardwood, or a softwood according its density, willow may, according to the cut used, be either. So, when made from heartwood, bats were automatically 50% heavier than they are today for the same size of blade and, until the 1835 change limiting the length of the bat, could be far larger than a modern bat too. Similarly, as the limitation on the width of the bat only came into the 1774 laws, it is possible that some bats of the epoch would have been even heavier and more unwieldy than the bludgeons of the 19th Century if they were larger than the maximum size that was permitted post-1774.

There is no requirement that bats should be made of willow rather than an alternative wood, but its light weight, strength and soft, spongy structure, added to its attractive cosmetic appearance make it ideal. The sponginess comes from the cellular structure of the wood. Whereas hardwood has mainly fibres, with thick cell walls, softwood is almost exclusively composed of *tracheids*, with thin cell walls, while fibres and vessels are absent.

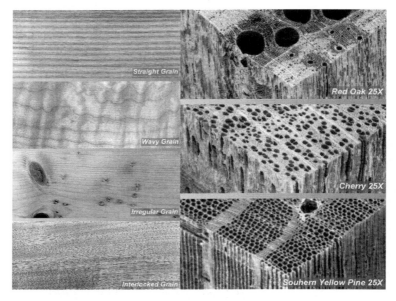

Examples of grain structure to the naked eye and under the microscope. The difference between the interlocked grain of willow (bottom left) and other types of grain is obvious. On the right we see the grain structure of hard woods under a microscope: some hard woods have large vessel elements that form visible pores in the wood that form so-called "open grain" (top); heartwood willow is similar to cherry (centre) with closed grain and small vessels. Softwoods such as sapwood willow have no vessel structure and are defined as "neither open nor closed" (e.g. Pine, bottom, which has a few channels that carry resin, but nothing like the wide vessels of Red Oak).

Wood has various physical properties that distinguish it from metal or plastic: in particular, when new it is humid and it shrinks as it dries, but it does not shrink evenly in all three axes. Bat willow is cut from the sapwood and thus has much higher water content than heartwood. Approximately 72% of the total water content is free water – sap – while the remaining 28%

is bound water that permeates the fibres of the cell walls. As the willow dries, the sap is eliminated first from the green wood, without changing its physical dimensions. Once the free water has dried, the bound water also dries, causing the cell walls and hence the wood itself to shrink, however it will do this in to a different degree along each of the three axes. Longitudinally, that is, along the grain, willow is no different to other woods in that it shrinks very little. A 1 metre block of willow will shrink by just 0.1mm (0.01%) along its length on drying. In contrast, willow is one of the types of wood that has a considerable difference in its radial and tangential shrinkage.

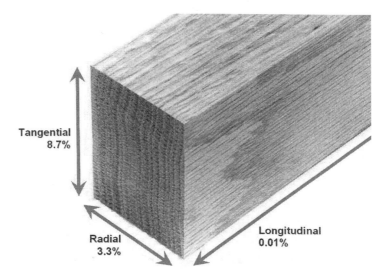

Willow shrinkage on drying. When freshly cut, the wood is saturated with sap – this is green wood. The sap accounts for about 72% of the total water content. The rest of the humidity is bound water that saturates the fibres in the cell walls. As the bound water dries, the wood shrinks by different amounts in different directions, leading potentially to warping and bending. To avoid this, the wood must be cut perpendicular to the grain, with the grain aligned as linearly as possible along the tangential direction so that it retains its original shape.

The ratio of tangential to radial shrinkage on drying for willow is 2.4 (8.7% tangentially, 3.3% radially), making willow much more susceptible to warping and bending as it dries than other woods such as hickory, sycamore or walnut for which the ratio is much smaller. To ensure that willow retains its shape as it dries, it must be cut so that the ring pattern runs across the face of the cut, with the long axis orientated towards the centre of the trunk in the radial direction. The closer the grain is to being parallel straight lines in the tangential direction, the better the piece of wood will retain its shape as it dries. The final water content of the dried wood depends on the relative humidity; in humid conditions wood will re-absorb water into the cell walls, swelling as it does so and increasing in weight. However, even if wood is completely soaked in a water bath, the maximum water content will be only 28%. Even so, this property of wood to absorb atmospheric humidity means that bats must be treated to avoid swelling and increase of density given that the moisture content of willow will typically change by 1% for each 5% change in relative humidity in the local atmosphere up to 80%. For relative humidity above 80%, which can be found in the tropics, the rate of increase of moisture content is much faster. Traditionally linseed oil has been used to create an impermeable outer protective layer for the bat to keep out humidity; many batsmen also use a protective plastic guard on the vulnerable toe of the bat to stop water seeping in from damp pitches.

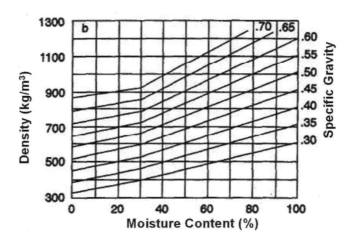

The relationship between the density of wood and the moisture content for a range of specific gravity values for raw, green wood. As wood shrinks as it dries, the relationship between the density and the moisture content is not a simple linear one. Adapted from "Specific gravity, moisture content, and density relationship for wood", Simpson (1993), United States Department of Agriculture Forest Service, Forest Products Laboratory General Technical Report FPL-GTR-76.

As a result of wood's shrinkage as it dries, it has some non-intuitive physical properties. The relationship between volume and mass is normally a simple one:

$$\rho = V/m$$

Where ρ is the density, V the volume and m the mass.

If you add a higher density fluid – humidity – to wood, you will increase its density but, as the wood swells when water is added, the volume will increase, so the density will increase in a non-linear fashion. The density varies with the moisture content and the specific gravity of the raw, green wood as:

$$\rho = \rho_w G_b \frac{(1 + M/100)}{1 - S}$$

Where ρ_w is the density of water, G_b is the specific gravity of the green wood, M is the moisture content by volume as a percentage and S is the volumetric correction term.

The volumetric shrinkage correction term varies between species, but can be well approximated[13] by

$$S_t = 0.265 \ G_b$$

Thus, volume corrected, the density varies with moisture content according to the following relationship

$$\rho = \frac{G_b(1 + M/100)}{1 - 0.265aG_b}$$

A reference moisture content of 30% is used for the volume correction term so, for other moisture content levels, we include the corrective term "a" in the equation to account for it, which is defined as follows

$$a = (30 - M)/30$$

What is the importance of this density relationship for cricketers?

The bottom line is that, logically enough, a piece of dry wood is lighter than a piece of damp wood. This concerns the bat manufacturer who needs to produce a specific weight of bat from the raw block of wood and concerns the batsman who will be concerned by the balance of the bat.

- If the raw block of wood dries insufficiently before being fashioned, the bat will be heavier than expected for the same volume of wood. Later treatment of the bat (e.g. oiling or putting a protective plastic covering on the blade) will seal in this unwanted humidity. To get a good, lightweight blade, it is essential to dry the wood well.

- If the raw block of wood dries unevenly, there will be density variations within the bat. This will change its balance and will lead to the bat not feeling right in the batsman's hands. This may, in turn, lead to difficulties for the batsman in executing his favourite shots with the expected precision.

Thus, particularly for the highest quality of bat, designed for international use, the drying of the raw wood and a rigorous quality control of the drying process is essential. This favours an extremely slow drying technique being used. Usually the trunk will be cut into pieces of equal length close to the required length of the blade of the bat and then cut into individual pieces called clefts that approximate to the shape of a bat blade. Only then will the drying process start. A major bat manufacturer such as British company Gray-Nicolls will use about 1200 trees per year[14], so this is no trivial operation.

The purpose of the bat is to apply impulse to the ball and to set it in movement in a controlled fashion in the batsman's chosen direction. Since the 1960s, bat weights have tended to increase steadily with the aim of applying more power to the ball when it is hit and, in particular, to clear both the in-field and the boundary rope more easily. We can see the effect this has had by looking at some bat weights used by some of the top players of past and present and looking at their Test Six-hitting.

	Career	Bat weight	Test innings/Six hit
Ian Botham	1977-1992	3lb 8oz	2.4
Clive Lloyd	1966-1985	3lb 2oz	2.5
Aravinda de Silva	1984-2002	3lb	3.3
Sachin Tendulkar	1989-xxxx	3lb	4.5
Steve Waugh	1985-2004	2lb 8oz	13
Don Bradman	1928-1948	2lb 2oz	13
Jack Hobbs	1908-1930	2lb 2oz	13
Alistair Cook	2006-xxxx	2lb 8oz	23

Do heavier bats really mean bigger hitting? One indicator is six hitting in Tests over a career for batsmen. Although not all batsman are equally inclined to take the aerial route to the boundary (see Alistair Cook!), overall there is a good correlation between the bat weight and six-hitting feats for this fairly random sample of batsmen[15]. Although the batsman's style of play has an obvious influence (famously, after getting out hooking several times in the Caribbean in 1974, Geoff Boycott eliminated this shot completely and never again hit a Six in a Test), we can see that this is a not unexpected correlation.

While using a heavier bat allows more power to be applied to the ball, it also requires the batsman to do more work to move the bat at a given velocity. There is though, a much more serious problem with heavy bats. Shots are essentially pendular with the bat rotating along an arc with the batsman's shoulders as its pivot. The kinetic energy of such a rotating system is

$$E = \tfrac{1}{2}\, \omega^2\, mL^2$$

Where ω is the angular velocity in radians per second, "m" is the mass and "L" is the distance to the pivot.

This is the moment of rotational inertial. Obviously, the greater the mass of the bat and the greater the distance from the pivot, the greater is its moment of inertia and the greater the force that will be applied to the ball on impact, hence some batsmen have preferred to increase the effective distance from the pivot by using a long-handled bat. However, no one gets anything free of charge in physics. The cost of a larger moment of inertia of the bat is that, once moving in a given direction, it is harder to stop it moving in that direction – this is what inertia means in practice... like a charging rhinoceros, the bat will be slow to change course if the ball is not where the batsman expects it to be when the stroke is executed and he attempts to correct his shot. Once the batsman is committed to hit the ball, it is harder to adjust for lateral movement with a heavier bat and, when a full swing of the bat is used, adjustment is almost impossible. This makes edged shots more likely so, to compensate, the edges of the bat are reinforced so that when the edge is taken there will still be a significant mass of wood behind the shot[16].

The impact between bat and ball is a complex system. The bat and ball are moving in opposite directions at different velocities. The impact is brief, lasting less than one millisecond and the force applied to the ball is a function of time and approximates to a sine squared curve, with the maximum applied force approximately at mid-impact. The ball behaves in a non-linear way on impact, as it will become stiffer the more that it deforms as the impact progresses, while the bat itself is elastic, but has a decreasing coefficient of restitution at greater speeds of impact of the ball.

If v_1 and v_2 are the ball and the bat's velocity respectively with "a" the velocity before impact and "b" the velocity after impact, the bat's coefficient of restitution can be defined as

$$e = \frac{v_{1a} - v_{2a}}{v_{1b} - v_{2b}}$$

If we assume that momentum is conserved in the impact, the ball's exit velocity is given by

$$v_{12} = \frac{(m_1 - em_2)v_{1b} + (m_2 + em_2)v_{2b}}{m_1 - m_2}$$

Where m_1 and m_2 are the masses of the ball and of the bat respectively.

For a good quality bat of English willow, we can approximate the mechanical and physical properties of the bat by the following[17]

Density	450 kg/m³
Elastic modulus	9.8×10^9 Nm⁻²
Shear modulus	6.7×10^9 Nm⁻²

To model the impact between bat and ball we need a model for the ball's behaviour as it deforms too. The composite composition of the ball makes this a non-linear function. However, the ball's shear modulus can be approximated by

$$G(t) = G_\alpha + (G_0 - G_\alpha)\, e^{-\beta}$$

Where β is a visco-elastic material constant.

Using this time-variant model and a constant Bulk Modulus for the ball, the ball's mechanical properties can be approximated as

Bulk modulus	69×10^6 Nm^{-2}
G_0	41×10^6 Nm^{-2}
G_α	11×10^6 Nm^{-2}
β	9000

Bat manufacturers and players will talk about the *sweet spot* of the bat: the impact point where shots are cleanest and most effective because the ball exits with the greatest velocity. The sweet spot is determined by the vibrational characteristics of the bat on impact with the ball considering that the bat is fixed at one end – the batsman's grip – and free at the other.

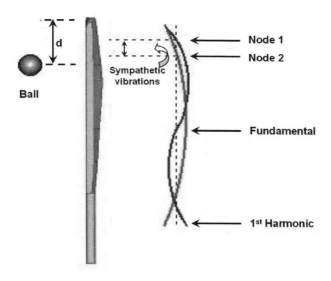

The position of the sweet spot of the bat is determined by its vibrational modes on impact with the ball. The sweet spot occurs between the nodes of the fundamental frequency and the first harmonic at the free end of the bat – the toe end. In between these two nodes the ball exits the bat with the highest velocity.

The parameters of the bat and ball described above can be used analytically to calculate the ball exit velocity v_{12} from the relations above for a middled shot. Unsurprisingly, the ball's exit velocity is not simply the sum of the bat velocity and the ball velocity, but instead is always lower than this sum and the exit velocity depends, sometimes strongly, on the exact point of impact on the bat. Modelling by Vaggeeram Hariharan and P.S.S. Srinivasan using an engineering software code[18] showed the for different combinations of bat and ball speed the highest exit velocity of the ball did indeed fall in the sweet spot between the node of the bat's fundamental and 1st harmonic vibration mode. Bat velocity was found to be more important than ball velocity in determining the speed of the ball off the bat. However, a caveat in this study is that the ball speeds used were unrealistically high as they correspond to the speed out of the bowler's hand, not the speed of the ball off the pitch.

In the regime of lower ball and bat speeds, we can approximate the problem to one where the ball exit velocity is directly related to the bat's effective coefficient of restitution "E".

$$v_{12} = v + v_2 \cdot (1 - E)$$

Where "v" is the velocity of the ball off a dead bat.

In this simplified formulation, the coefficient of restitution depends on the point of impact and hence the mass of wood behind the impact point. It varies from approximately 0.1 near the tip of the bat to approximately 0.3 in the centre of the blade, so a value of the order of 0.2 in the sweet spot is quite close to the truth. At the same time, for a full swing of the bat, the bat's angular velocity will vary somewhat along the length of the blade, being greatest at the tip.

Curiously, the velocity of a ball off a dead bat is, in reality, independent of the speed of the ball on impact. Similarly, because the ball's contact with the bat is so brief, the firmness of the batsman's grip on the bat handle does not influence the rebound either because the time taken for the shock wave of impact to travel up the blade to the point of grip and back again is about twice as long as the time that the bat and ball are in contact.

We can say that

$$v \approx 6 \text{ m/s}$$

So, if

$$E=0.2$$

$$v_{12} = 6 + 0.8 \, v_2$$

For a full-blooded shot where the batsman swings straight through the ball, with the point of contact of the bat moving at 25 m/s

$$v_{12} = 26 \text{ m/s}$$

As we know that the distance travelled is maximised when the angle of elevation of the shot is 45° and that the horizontal component of velocity is v_{12} sin 45, a blow hit at this velocity will travel 35 metres before pitching. To clear the boundary, the bat must be swung at a velocity close to 30 m/s.

The vibrational modes of the bat and hence the location of the sweet spot can be changed by changing the weight distribution of the bat and, to a lesser degree, by changing the springing in the handle. A batsman who gets on the front foot a lot and drives will generally prefer a bat with a low sweet spot so that shots closer to the toe of the bat are hit with power. In contrast, a back foot player would be advised to use a bat with a higher sweet spot.

As we can see, a bat may *look* like a simple piece of wood, but there is actually a lot of quite complex physics behind it. The bat maker's art is to take that simple piece of wood and, using traditional arts, tune it so it vibrates exactly in the required manner like a fine musical instrument as the batsman plays. He must adjust the balance refine the weight and its distribution and, in general, attend to his client's every whim.

Endnotes

[1]   One of the striking things about these laws is their remarkable lack of definition because attempts to bend the laws and go against their spirit have been surprisingly rare. Note the use of the subjective and highly ambiguous word "unacceptable" in a fundamental part of the definition. Cricket still has a touching innocence in some respects, which assumes that the unscrupulous would not seek to take advantage of such ambiguities in the laws. Even today you will sometimes hear the phrase "it's just not cricket" to describe sharp practice in some area of life.

[2]   A million Harry Potter fans will foam at the mouth and point out that, in Quiddich, you may use your broom both to shoot and to block shots and that the Beaters use bats. True, but Quiddick is mainly a hand-to-hand passing and shooting game and the best Chasers, such as Ginny Weasley, score goals throwing the Quaffle. Shooting with the brush end of the broom is a complicated manoeuvre and usually only done to show off.

[3]   The incident happened at the end of the first innings of the 1st Test at Perth, shortly before the close of play on Day 1 (December 14th 1979). Australia had recovered from 20-3 to 219-7 when Dennis Lillee came out to bat with his aluminium bat. After facing three balls without incident, it was the fourth ball of his innings that caused the controversy. You can see film of this incident and the protests by England captain Mike Brearley at the URL: http://www.youtube.com/watch?v=7Pak_0L3rhc. A furious Denis Lillee took dreadful revenge on the England batsmen next morning when he had a ball in his hand, leaving England reeling at 14-3, just thirty minutes into their innings, with both openers (Randall and Boycott) dismissed for ducks. It may not be entirely irrelevant that he had a commercial interest in marketing the aluminium bat as an alternative to the traditional wooden one. It must be stressed that, at the time of this incident the aluminium bat was perfectly legal, even if both captains objected to its use, although for different reasons.

[4]   The incident is shrouded in some mystery, but caused enormous shock waves in the game. It is often stated that the player in question was Thomas "Shock" White, but this seems to be a totally different and blameless individual, who may not even have been called Thomas! It seems that the match in question was Chertsey v Hambledon, which was effectively a Surrey v Hampshire fixture, with each club being allowed to field four guests. The game was played, at Chertsey, on September 23rd & 24th 1771, for the sum of 50 Pounds (equivalent to about 70 000 Pounds in 2010 terms) and was won by Hambledon

"only by a single notch" according to the newspaper account (i.e. by a single run), having aggregated 218 in their two innings.

Thomas White came out to bat on the first day of the match with a bat as wide as the stumps themselves. Whether he was exploiting a loophole in the laws set down in 1744, in the hope of winning the match, or just making a protest about a lack of standardisation of bat sizes is not known. However, by making Hambledon, the MCC of the day and the effective guardians of the laws of the game of cricket, the victims, White provoked the rule change. You can read a lot more about this incident in a fascinating piece at the URL: http://www.jl.sl.btinternet.co.uk/stampsite/cricket/histories/monster.html.

5   This has led to some remarkable incidents when players have used non-standard bats in First Class matches. Phil Edmonds, no mean batsman, particularly earlier in his career (at least, when the mood took him) on one occasion went out to bat for Middlesex with a completely unshaped piece of wood as a bat – i.e. the original cleft from which a bat is pressed – with which he proceeded to savage the bowling. A less happy story was of another, well known Lancashire cricketer, who decided to go out to bat in a match with a 2-inch (5cm) wide practice bat (that it, just the centre of the bat, without the edges) that is used to check in practice whether or not a player is middling the ball correctly. Needless to say, this peculiar experiment did not last long and the player's teammates were not particularly impressed when his innings ended rapidly.

6   Each time I go to Patrick Moore's house in Selsey (Sussex) I admire the old bat – presumably 1940s vintage – that he has resting against the fireplace in his study. It has evidently seen a lot of use but, compared to modern bats, looks like a child's toy because it is so small and light.

7   Almost his entire career he was England's rival to Don Bradman and eclipsed by him, Hammond's scores would have been regarded as quite sensational against any other yardstick. His average is one of the highest ever, despite not recording a century between April 1933 and July 1935, during a prolonged dip in form and despite an unhappy return to Test cricket after the war when he averaged just 30 over 8 Tests when well over forty years old. At the end of the final Test before World War 2, which ended at the Kennington Oval (London) on August 22nd 1939 (this was a 3-day Test, as was the case for all matches against the newest Test teams), he averaged 61.45.

Among Hammond's feats are a sequence of 251, 200, 32, 119* and 177 over three Tests in the 1928/29 Ashes series in Australia, He also ended the Bodyline tour of Australia and

New Zealand in 1932/33 with 101 and 75* in the Final Test v Australia, followed by 227 and 336* against the suffering New Zealanders. No less than five times he managed a sequence of three centuries in four Tests.

[8] This is particularly true as he played 37 of his 52 Tests against England, who were far and away the strongest rivals for Australia. South Africa were extremely weak at the time, the West Indies (and New Zealand – who did not play Australia until much later) were new to Test cricket and even India were far from the majestic team of recent times.

[9] As bat weights are traditionally given in pounds and ounces, the metric equivalent here is approximated to the nearest 50g (1.8oz).

[10] Some bats may be made from Salix terasperma – the Indian Willow – this wood though is regarded as being of inferior quality in bat-making and so is generally used only for bats at lower levels of the game.

One of the less-known uses of willow is that its bark contains salicin, closely related to salicylic acid – the active ingredient of aspirin. Aspirin was discovered thanks to the separation of salicin crystals from willow bark.

[11] http://www.engineeringtoolbox.com/wood-density-d_40.html

[12] http://workshopcompanion.com/KnowHow/Wood/Hardwoods_&_Softwoods/2_Mechanical_Properties/Mechanical_Properties_Table_1.htm#Willow

However, as the density depends strongly on the moisture content, different sources will quote slightly different values according to the way that the measurement is defined.

[13] Stamm, A.J.: 1964, Wood and cellulose science. The Ronald Press (New York), p. 549

[14] http://www.telegraph.co.uk/finance/newsbysector/industry/8222896/Batmaker-Gray-Nicolls-goes-to-the-crease-in-the-Ashes.html.
The article also contains some interesting background information on the company, how it operates and bat manufacture in general.

[15] The list was chosen to include a range of batsmen past and present for whom their normal bat weight is reasonably well documented. The Six-hitting statistics were only calculated *after* the list was formulated. I had expected to see some kind of correlation, but not one this good.

16   This was the reasoning behind Ian Botham's play in his famous 149* at Headingley in 1981. He reasoned that with a heavy bat with very fat edges, a lot of close fielders and a full swing of the bat, even shots that were not centred would be likely to clear the field and go for boundaries. 148 balls and 219 minutes later, 27x4 and 1x6 spectacularly proved him right. 76.5% of his runs came from boundaries.

17   Taken from Hariharan and Srinivasan, 1995, "Inertial and vibrational characteristics of a cricket bat" (published on vibrationdata.com) [http://www.vibrationdata.com/cricket_bat.pdf].

18   See Hariharan and Srinivasan, 1995, "Inertial and vibrational characteristics of a cricket bat" (published on vibrationdata.com) [http://www.vibrationdata.com/cricket_bat.pdf]. Most numerical studies of this problem have dealt exclusively with baseball bats rather than cricket bats.

CHAPTER 8

PAKISTANI HYDRODYNAMICS

"We won fair and square"

Waqar Younis, Lords, August 23rd 1992

August 9th 1992. Kennington Oval (London). Fourth morning. England, facing a first innings deficit of 173, are seemingly hauling themselves slowly to a position that offers some hope in the deciding Test of the summer that is becoming steadily more fractious. Robin Smith and Chris Lewis have put on 61 against an increasingly old and soft ball, reducing the arrears to 20. With Derek Pringle and Neil Mallender, both with First Class centuries to their name, still to bat there is a chance that England can eke out some kind of lead. In the 2nd Test at Lords, with Pakistan chasing a small target to win, only a remarkable 9th wicket partnership between Wasim Akram and Waqar Younis who had come together with Pakistan 95-8 had saved them from a demoralising defeat. There is a feeling that a lead of 120 might be just enough to see another Pakistan panic set in. Chris Lewis, not normally a man to hang around, has accompanied Robin Smith for just under two hours, scoring only 14 off 82 balls as the Pakistan bowling is gradually worn down. The new ball is not due for another twenty overs and if this pair can survive until then, a miracle might just be possible. Mushtaq Ahmed has taken fourteen wickets so far in the series with his leg breaks and googlies and, although he has not been destructive, his bowling has proved somewhat less easy to read than Linear-B to some of the England batsmen. Now he strikes a deadly blow, tempting Chris Lewis out of his crease to provide debutant wicket-keeper Rashid Latif with his first Test stumping. With an end open, Wasim and Waqar suddenly start to make the ball talk, swinging it wildly. After surviving nine deliveries, the tenth, bowled by Wasim, crashes into Derek Pringle's stumps. All Robin Smith's efforts to farm the strike and protect the tail are now hopeless as Mallender and Tufnell fall to successive balls, Tufnell becoming the eleventh England batsman of the match to fall without the intervention of a fielder as his stumps are scattered too[1]. Devon Malcolm delays the inevitable for a short while but, finally, a yorker from Waqar crashes into his stumps. England have slumped from 159-5 to 174 all out. Pakistan need just 2 runs to win the Test and the series.

Through the series, Wasim and Waqar have taken a combined 43 wickets, of which a remarkable 17 have been bowled, with another 8 LBW. In the last two Tests, 17 of 23 dismissals by Wasim and Waqar have been bowled or LBW. In contrast, England's spearhead, Devon Malcolm, has managed just 3 of his 13 dismissals without the aid of a fielder. Something is amiss.

Why is the old ball proving to be so much deadlier in the hands of the Pakistani pace attack than the new ball[2]? Why were so many England batsmen being caught in front of the stumps or, worse, missing straight balls?

Five years after the Mike Gatting – Shakoor Rana incident, feelings were still pretty raw between the two sides. All summer there had been whispers of ball tampering: that the Pakistan bowlers were using illicit or underhand means to damage the ball and gain an unfair advantage. After seeing his batsmen collapse horribly twice in the Test facing Wasim and Waqar, England manager Mickey Stewart alluded darkly at the end of the match that he knew how the bowlers had managed to make the old ball swing so violently but he was not going to reveal the secret of their success. What no one was prepared to accept was that the Pakistan attack had discovered a radically new method of bowling the ball which, in the hands of a pair of supremely talented fast bowlers, posed a problem that the England batsmen were simply not capable of resolving. Pride would not permit the team and its management to accept that they had been outthought and outplayed. Reverse swing had arrived.

Unfortunately, as has so often happened in Pakistan cricket, two particular incidents in the series distracted attention from their amazing success and set them up for accusations of cheating. Crucially, it laid the ground for a far more serious affair in 2006[3]. Due to these incidents that produced ready-made reasons to doubt that their success had been achieved fairly, the bowlers did not get the credit that they were due for their bowling. First, during the Lords Test, the camera had caught Aqib Javed in close up, working the ball with his thumb nail. Richie Benaud, a neutral commentator, famously gasped "hey up, steady on!" Later, ironically after the Lords Test furore, during the Lords ODI, with the match struggling through a second morning due to rain, England had reached 140-5 at lunch, chasing 205 to win. When the ball was returned to umpires Ken Palmer and John Hampshire they saw that it had been deeply gouged. Match referee Deryck Murray was summoned and agreed that the ball had been deliberately slit by person or persons unknown. Intikab Alam, the Pakistan team manager was called to the dressing room too and informed that under the previsions of Law 42 the umpires were going to change the ball, a verdict that he accepted.

Play continued after lunch with a different ball and at 191-6, with just 13 more required and time enough to get them without taking risks, an England win looked a formality. Captain, Javed Miandad, tossed the undamaged replacement ball to Waqar and Wasim for the closing overs. Suddenly the ball started to swing. Waqar whistled a ball through Richard Blakey's defences when the young Yorkshireman, playing only his second ODI had seemed impregnable. Wasim then removed Lewis and de Freitas in the same over and England were 193-9. With just 4 needed to win, Waqar bowled Richard Illingworth (who, with four First Class centuries to his name, was one of the best number eleven batsmen that England has played in recent history) with the second ball of the final over of the match to complete a stunning turnaround.

When the news broke that the ball had been changed during the England innings the talk was of a tainted victory, forgetting that the devastating spell of bowling that won the match had been achieved with an undamaged ball. When asked to comment, Waqar said, not unreasonably "I don't care what anyone thinks... the new ball swung more anyway. Every time we win people start saying these things. We won fair and square." Waqar and Wasim had shown that reverse swing could be obtained quite fairly, without tampering with the ball's condition, if you had the skill to do it. It would take England's bowlers thirteen years to master that skill.

So, what are swing and reverse swing and how are they obtained?

Unlike seam movement, which is obtained by the optical principle of reflection as the raised seam hits the pitch, swing is due to aerodynamic forces on the ball in the air. Powered, heavier than air flight has existed for over a century now since the Wright brothers' first flight at Kitty Hawk on December 17ᵗʰ 1903, so one would imagine that we would understand perfectly the principles of aerodynamics. Extending them to a simple cricket ball should be child's play.

In fact, it does not take long to discover that there is a great deal of argument, some of it quite bitter, even about how exactly an aircraft wing generates lift – although no one disputes that it does. In fact, some studies even have concluded that the swing of a cricket ball is nothing more than a simple optical illusion. Similarly, to the bewilderment of many, several laboratory studies showed that it is totally impossible for a cricket ball to swing, and even those studies that demonstrate that it can, have found that humidity has no effect either (any suggestion that the ball swings more in very cloudy conditions than when the sun is out tends, unhelpfully, to be rejected as unphysical).

The classical explanation of both how an aeroplane flies and how to make a cricket ball swings invokes *Bernoulli's Principle*, or the *Bernoulli Effect*. This was formulated by Daniel Bernoulli in 1738[4].

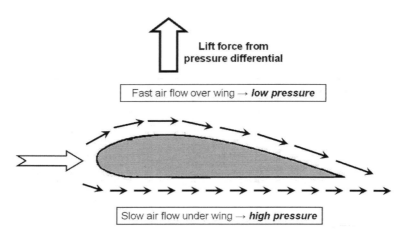

The conventional theory of lift being generated on an aircraft wing due to the pressure differential caused by the Bernoulli Effect. Air flows faster over the upper, curved surface than over the lower surface causing a pressure differential that pushes the aircraft upwards towards the region of lower pressure. Under this standard theory an aeroplane would be incapable of inverted flight. A variant of this theory was accepted for many years as the source of the swing of a cricket ball.

In fluid dynamics the Bernoulli Principle states that if a non-viscous fluid's speed increases, its pressure will decrease. If we have fast-moving air above the wing's surface and slower-moving air underneath the pressure differential will exert a lift force. The lower pressure over the curved top of the wing can be observed by watching condensation forming over the wing surface in very humid air as the pressure and hence temperature drop.

The theory is identical for a cricket ball: make the air flow faster over one side of the cricket ball than over the other and a pressure differential will develop that forces the ball to swing through the air. Bowlers have reasoned that if you can make one side of the ball heavier, for example, by soaking it with sweat, it will move slower through the air and thus swing towards the side that is heavier[5]. This is similar reasoning to refraction of light: if you shine a beam of light at an angle through a medium – for example, a tank of water – the speed of light will be much slower in the denser water than it is in the air, so the beam will turn sharply towards the denser medium[6].

However, there is a problem. If the Bernoulli Principle were the principal driver of lift, an aeroplane would plunge out of the sky when placed in inverted flight. Many pilots will tell you though that they have flown inverted and not fallen out of the sky. Similarly, a symmetrical wing profile, or a profile in which the lower wing surface has more curvature than the upper surface, such as the Whitcomb Supercritical Foil would not work[7] and a nice, symmetrical cricket ball could not possibly swing. In fact, to generate enough lift to support a wide-bodied jet through just the Bernoulli Effect, the wing would have to be so thick and so strongly curved that the drag at flying speeds would overcome lift, with unfortunate consequences for all those on board. Bernoulli alone is not enough to explain the lift generated by a wing and experts certainly do agree that it is not the reason why a cricket ball will swing. We must look elsewhere.

What about an alternative method of generating lift from an aircraft wing that is vigorously supported by some experts: Newtonian mechanics?

Let us start with Newton's First Law:

> *Every object in a state of uniform motion tends to remain in that state of motion unless an external force is applied to it.*

This law states simply and obviously that a cricket ball will not swing – implying lateral acceleration – unless a force is applied to it. We can estimate this force by applying Newton's Second Law:

The relationship between an object's mass m, *its acceleration* a, *and the applied force* F *is* F = ma.

Suppose a ball delivered at 120km/h swings 15cm away from the batsman in the air before pitching. What is the lateral force that is acting on the ball?

We have already seen that the ball will slow by about 12% before pitching, so its average speed through the air will be 113km/h. If it pitches about 6m in front of the batsman, it will travel 13.5m through the air and take 0.43 seconds to pitch.

If there is no initial lateral velocity

$$s = \tfrac{1}{2} at^2$$

Where "s" is the distance, "a" the acceleration and "t" the time. Thus

$$a = 2s/t^2$$

And the lateral force acting on the ball is

$$F = ma$$

$$F = 2ms/t^2$$

Where "m" is the mass of the ball (0.16kg)

$$F = 2 \times 0.16 \times 0.15/0.43^2 \text{ N}$$

$$= 0.26N$$

This does not *sound* like much, but it is, in fact, fully one sixth of the force that gravity exerts on the ball. In other words, the lateral force acting on the ball is substantial. Where could this force come from?

Newton's Third Law states:

For every action there is an equal and opposite reaction.

This law is typified by a rocket engine. The rocket burns fuel and pushes the exhaust gases out of the bottom of the rocket at very high speed. Although the rocket is much heavier than the exhaust gases, the impulse provided by the exhaust gases generates a substantial acceleration because the conservation of energy obliges it to accelerate in the opposite direction to the exhaust gases so that the net energy of the system is zero. A jet engine behaves in a similar way: hot exhaust gas is blown backwards, pushing the aircraft forwards.

How does this affect our aircraft wing? As air flows over the upper surface of the wing, viscosity will make it stick to the wing in the same way that a stream of water dribbled over a horizontal cylinder will stick to the curve of the cylinder – the *Coanda Effect*. As it leaves the trailing edge of the wing it will not travel horizontally (as in the Bernoulli explanation), but rather be forced downwards. The mass of air leaving the wing in a downwards direction will cause an upwards force to be exerted on the wing, giving it lift.

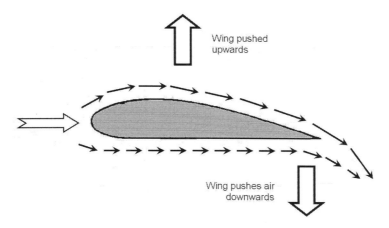

A Newtonian explanation of lift. The wing does work by pushing air downwards as it leaves the trailing edge. By Newton's Third Law, the downwards force of the air exerts an equal and opposite upwards lift force on the aircraft.

Under Newton's Third Law

$$E = \tfrac{1}{2} mv^2$$

Where "m" is the mass displaced and "v" the velocity at which it is displaced.

Air has a low density, so a huge volume of air must be displaced at high velocity to exert a significant force against a large mass.

The density of air varies a great deal with temperature and pressure. Hot air is less dense, which is why a hot air balloon rises. There is a quite significant difference – just about 10% - in the density of the air between a hot day in Mumbai and an October morning in Hamilton (New Zealand). Let us take a warm July afternoon at Lords as our standard and see what mass of air the ball will displace before pitching. The density of dry air at 25°C is 1.18kg/m³. If we take a ball of mid-range circumference (22.7cm), its radius and thus cross-sectional area will be:

$$r = 22.7/2\pi$$

$$A = \pi r^2$$

$$= 22.7^2/4\pi \text{ cm}^2$$

$$= 41 \text{cm}^2$$

And the volume of air displaced will be

$$V = A \cdot l$$

Where "l" is the distance of flight to pitching. For the example that we have given above, l = 13.5m, so

$$V = 0.055 \text{ m}^3$$

Thus the mass of air displaced is

$$M_{air} = \rho V = 1.18 \cdot 0.055 \text{ kg}$$

$$= 0.065 \text{ kg}$$

As the mass of the ball is ≈ 0.16 kg, it displaces about one third of its own mass in air in flight. However, we have already seen that the lateral force on the ball is equivalent to about one sixth of its mass so, even if we position the seam of the ball in such a way that it deflects air efficiently laterally – about 20° off the direction of flight is generally reckoned to be the best seam orientation to get the best swing – we either need to deviate the displaced air through a large angle to get the required lateral thrust, or somehow accelerate considerably a smaller amount of displaced air laterally, neither or which is terribly plausible (no one has yet accused a bowler of smuggling a ball onto the field of play with a small, incorporated jet engine to make it swing more). Half of all the displaced air would need to leave the ball in a direction orthogonal to flight to achieve even a relatively modest 15cm of swing!

However, we should not forget that the pressure on the surface of a cricket ball is considerable. Normal atmospheric pressure is a small matter of 1.017kg/cm^2. For a ball of 22.7cm circumference, the surface area is 164 cm^2, so the atmospheric pressure acting on the ball is equivalent to 167kg – close to the mass of two fast bowlers! So, 1640N of atmospheric pressure acts on the ball, 820N on each face and we need only an atmospheric pressure difference of 0.26N between the two faces – less than one part in one thousand – to make the ball swing. How can we achieve that pressure difference?

Remarkably, writing as far back as 1958 in his seminal tome "The Art of Cricket", Sir Donald Bradman, helped by Prof. Raymond Lyttleton of St. John's College, Cambridge, came up with a theory that is close to what is accepted today.

Bradman's three key factors to obtain swing were:

1. A new ball with a shiny surface.
2. A humid atmosphere, with cloud.
3. A wind blowing from the right quarter.

Of these, point (3) imposes the least strain on physics given that a wind in the right quarter will, logically enough, always blow the ball off course whatever other factors are involved, although for Bradman the ideal wind direction to help the swing would be at forty-five degrees to the direction of flight: i.e. coming from Third Slip to help the inswinger and from between the wicket-keeper and Square Leg for the outswinger. Similarly, Don Bradman points out that an express pace bowler – he cites the example of Frank Tyson – will always get the ball to swing less than a fast medium bowler such as Alec Bedser, although at the fastest pace it takes much less swing to beat a batsman than at medium pace because the batsman has so much less time to react and compensate[8].

Among the other factors that Bradman noted was that the seam appeared to act as a rudder. If the seam points towards the slips on release, it will swing away from the bat; if it points towards fine leg the result will be an inswinger. The biggest single factor though for him was the presence of a good shine on one side of the ball, while the other side had to be roughened-up.

With a smooth, symmetrical ball, the airflow effectively clings to the ball giving a smooth boundary layer with laminar flow at rest relative to the ball. According to the speed of the ball through the air this boundary layer will peel off symmetrically into a turbulent, low pressure wake at some point on the opposite side of the ball to the direction of motion. If the ball is delivered with an inclined seam, the point at which the boundary layer peels off will not be symmetrical and a pressure differential will be induced: the rough seam will hold the boundary layer onto the ball for a larger fraction of the circumference than on the opposite side of the ball, creating a lower pressure on the side that that the seam is facing[9]. This pressure differential is what generates normal swing. The greater the difference in pressure between the two sides of the ball, the greater the degree of lateral force and thus swing.

Detailed experiments have established the essential veracity of the Lyttleton model of normal swing. In 1983 the prestigious journal Nature published a study by Rabindra Mehta, who now works for NASA, of the behaviour of a selection of Reader balls to establish what parameters determine the degree of swing[10]. The biggest new issue was the degree of spin on the ball. It was found that the degree of backspin imparted to the ball on delivery was extremely important as, provided that it is imparted along the seam, it stabilised the position of the seam in flight. If the rate of spin was too high, the pressure difference was found to be reduced because the effect of the roughness was smoothed out. The spin torque must be greater than the torque from the pressure differential between the two sides of the ball.

The amount of pressure differential reaches a maximum at 105km/h (65 mph) at which point the seam side of the ball shows fully turbulent flow, while the smooth side of the ball shows laminar flow. This is the "banana swing" regime exploited by medium and slow medium pace bowlers. At slower speeds it is found necessary to increase the angle of the seam to maximise the degree of swing obtained; a club bowler who delivers at a gentle medium pace would be advised to turn the seam to around 30 degrees from the direction of flight and will get more swing if the delivery action imparts a much higher degree of backspin on the ball. There is one exception to the rule of maximum swing being achieved at 105km/h (29.2m/s): with the seam angled at 30° and only a small degree of backspin on delivery, the bowler will actually achieve a higher degree of side force on the ball at 150km/h (41.6m/s) than at 105km/h (29.2m/s). Unfortunately though it is extremely difficult for the bowler to control the amount of backspin that he puts on the ball at delivery. The fact though that obtaining an optimum degree of backspin stabilising the seam position and an optimum seam angle together can produce a much greater degree of lateral force on the ball could help to explain the phenomenon of the "jaffa" – the sudden, unexpected, devastating delivery from a fast bowler that appears out of the blue with no apparent explanation.

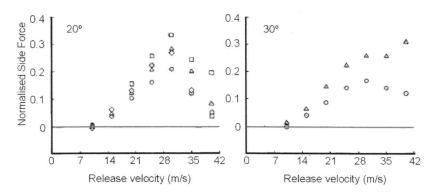

Normalised lateral force on the ball as a function of release velocity for seam angles of 20° (left) and 30° (right). The degree of backspin on release is denoted by: circles (5 rev/s); triangles (9.1 revs/s); squares (11.4 revs/s); diamonds (14.2 revs/s). For all seam orientations the degree of lateral force peaks sharply at around 30m/s in all cases except for a delivery with a seam angle of 30° and a backspin very close to 9 revs/s, for which the lateral force increases steadily with velocity – this is the "jaffa" delivery that requires very exact initial conditions to be possible. Adapted from Mehta et al. (1983, Nature, 303, 787-788).

In contrast, a new ball with a straight seam will barely swing, if at all. The experiments found that there was a very small lateral force at around 30m/s, even with a straight seam, but not enough to be useful to the bowler.

The relationship between the backspin required to stabilise the seam position and the degree of swing has led some people to confuse the mechanism for producing swing with the Magnus Effect, which applies to a spinning ball. We will meet the Magnus Effect later as the explanation why a ball that is spinning like a top will follow a curved trajectory in the opposite direction to the spin. However, the Magnus effect is one of the ways that spinners weave their magic, not seamers.

One controversial aspect of studies of swing bowling is the failure of any study to reveal any correlation of swing with humidity. Point (2) of Bradman's list above to obtain swing is cloudy and humid conditions. On many occasions it has been noted how the ball has swung violently for one side under lowering skies, only for the sun to break through when the opposition bats and conditions to become batsman-friendly, with no swing at all. Could the increased degree of swing be due to the differing densities of dry and humid air? As we have seen, warm air is denser than cool air, while humid air is less dense than dry air. Why?

Humidity is water vapour: H_2O, molecular weight 18 (H=1, O=16)

Dry air is a mixture of about 80% nitrogen (molecular weight 28) and 20% oxygen (molecular weight 32).

So, the molecular weight of dry air is given by:

$$\rho_{mol} = 0.8 \times 28 + 0.2 \times 32 = 28.8 \ g/mole$$

If we add lighter water vapour molecules to the mix, we reduce the mean molecular weight and hence the density.

We can calculate the density of humid air from the relation:

$$\rho_{humid\ air} = \frac{p_d}{R_d \cdot T} + \frac{p_v}{R_v \cdot T}$$

Where

p_d = Partial pressure of dry air
R_d = Specific gas constant for dry air (287.058 J)
T = Temperature
p_v = Pressure of water vapour
R_v = Specific gas constant for water vapour (461.495 J)

The partial pressure of saturated air at temperature T can be estimated by

$$p(mb)_{sat} = 6.1078 \bullet 10^{\frac{7.5T - 2048.625}{T - 35.85}}$$

At 300K – a warm day at Lords – the partial pressure of water vapour in saturated air is 35mb, equivalent to less than 4% water vapour in the gas mix, giving a decrease in mean molecular mass to 28.5 g/mole: an almost negligible change. As a result, the atmospheric density drop is very small indeed in humid conditions, much less than the density decrease between a cold spring morning and a hot afternoon at Madras, yet a good quick bowler will get swing both in Madras and at Durham in May. The study in Nature looked at the behaviour of the sample of balls in controlled conditions of humidity, but found that the degree of lateral force measured did not change at all. One theory that has been suggested to link humidity and swing has been the hypothesis that humidity makes the stitching on the ball's seam swell, increasing the lateral force due to increased turbulence in the wake of the seam, but the Nature study found that not only did the results not change with high humidity, but there was no detectable change in the ball's appearance and the height of the stitching even in very humid air. The question of how cloud cover and humidity can help swing bowling is very much open, with no obvious physical answer; that though is no consolation to the batsman when the ball swings around corners under a murky grey Headingley sky.

Once the initial shine is off the ball, the standard trick of the bowlers to shine one side of the ball and allow the other to roughen. This allows smooth, laminar flow over one face of the ball and turbulent flow over the rough face. Here we go into two different regimes of swing bowling.

- In conventional, new ball swing bowling, as we have seen, we use the seam as a rudder. The ball swings in the direction that the seam points.

In contrast:

- In standard old ball swing bowling, the seam is kept straight and polished to give a rough and a smooth side. The ball swings towards the rough side.

- In reverse swing bowling, the old ball swings in the opposite sense to the new ball, although delivered in exactly the same way, with the same seam angle.

Both methods rely on keeping one side of the ball as smooth and polished as possible and the other as rough as possible, but differ in their use of the seam and of speed.

For standard swing the roughness of the ball creates all the turbulence that the bowler needs, so there is no necessity to use the seam too. Similarly, at high speeds the effect of the roughness is reduced because the speed of the ball through the air ends up smoothing the air flow, so it is an advantage to take the pace off the ball.

Here, the swing bowler has conflicting interests. If the outfield is lush and the batting gentle, the shine can stay on the new ball for quite some time. Once the ball starts to lose its shine though it can be in the bowler's interest to let it deteriorate as fast as possible rather than to make an effort to keep the shine on it. It is in the batsman's interest to batter the ball as best as possible to roughen-up both sides and get rid of the shine. David Lloyd, former England opener, used to say that it was an opening batsman's duty to locate a nice piece of concrete wall around the boundary and try to hit at that piece of concrete as often as possible in an attempt to wreck the new ball quickly. After 10 or 15 overs being battered by Virender Sehwag, especially in India where the outfield is often less lush and green, any ball will have completely lost its shine. In contrast, the bowler will want a controlled deterioration of the state of the ball. If it is not roughening fast enough after losing its initial shine, the fielders will often return the ball to the wicket-keeper on the bounce rather than on the full, to help roughen it.

With conventional swing, the batsman can read the swing easily out of the bowler's hand as he knows that the ball will swing away from the shiny side. This leads to bowlers often trying to hide the ball in their hand to disguise the swing. What happens though when you read the ball's orientation correctly on delivery and play for the outswinger, only to see the ball boomerang into your stumps? This is the devastating reverse swing that England encountered so unhappily in 1992.

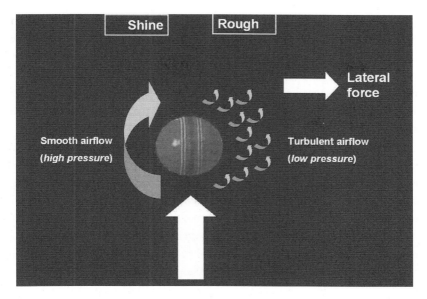

Making the old ball swing. The seam should be kept straight and one side of the ball polished up carefully. The airflow over the rough side will be turbulent and a pressure differential will be set up. Thus the ball will swing towards its rough side. This is conventional swing.

It happened that, in the early 1980s, Pakistan had a fast bowler who was also an expert exponent of swing. Imran Khan was a boyhood hero to a whole generation of cricketers around the world. His initial career gave little hint of what was to come and, after making his debut against England in June 1971, was in and out of the Pakistan side. By the end of 1976 he had played just 8 Tests and had only become a regular in the Test side in October 1975. His figures were not particularly impressive at that stage, averaging 23.2 with the bat and having taken 20 wickets at 42.6. His ninth Test, the traditional New Year's Day fixture at Melbourne was not an outstanding success either: 5 and 28 with the bat and 0-115 and 5-122 with the ball as Australia won a massive victory by 348 runs early on the fifth day. However, that second innings bowling spell was a warning to any Australian that if they thought that this guy Imran was a pushover, they had better start changing their ideas[11].

The bomb fell on an unprepared Australian side in the third and final Test at Sydney. 6-102 in the first innings, followed by 6-63 in the second and a century for Asif saw Australia barely avoiding an innings defeat. Imran Khan was so devastating that he bowled all but six overs from one end in the first innings and all but two in the second innings. The mighty Khan had arrived. After their struggles in the first two Tests, Pakistan had squared the series and cricket had a new star. His first nine Tests had produced 25 wickets at 43.5, his next nine produced 51 at 27.4 and his average would fall steadily thereafter.

As Imran Khan's career really took off he started to notice that something odd was happening. Holding the ball shiny side to the left to produce an inswinger, the ball was refusing to behave and was often producing a big outswinger instead. Oxford educated Imran, who single-handedly converted the University team from also-rans to a considerable force in one-day cricket, was also an intelligent and thinking cricketer and realised that he had found something important.

What Imran had discovered was a brilliant application of hydrodynamics, which he was able to teach his colleagues in the Pakistan team, leading to a generation of bowlers who time and again confounded their rivals. At high speeds – and Imran was a genuinely fast bowler – the airflow over the ball changes radically. When the seam is inclined, the airflow actually goes turbulent very quickly, even before it reaches the seam. The seam interferes rather than helping, pulling the airflow off the surface, separating it from the ball while the flow stays on the surface on the rough side and thus reversing the direction of the lateral force. This is reverse swing because now it is the shiny side of the ball that has the lower pressure and the ball will swing towards, not away from the shine. When the ball reverses, a batsman will instinctively play for an outswinger, only to discover the ball dipping in to him or vice versa. Why were so many England batsmen out bowled or lbw against Pakistan in 1992? They were tending to play around the ball: their instincts told them to expect the outswinger because they could see the shine on the inside, but the ball moved in the opposite direction, inside the bat.

A relatively new ball will only start to reverse swing at high speed, 140km/h and beyond, which is faster than most fast medium bowlers are capable of. However, as the condition of the ball deteriorates, reverse swing will appear at rather lower speeds. This explains the over-enthusiasm of some players for giving this deterioration a helping hand. However, as we have seen too, reverse swing can be obtained also quite legally. Techniques that help to roughen the ball without recourse to thumbnails, bottle tops and assorted abrasives are such things as always throwing the ball in from the outfield on the bounce, rather than letting the wicket keeper take it on the full.

Unfortunately though, as letting the condition of the ball deteriorate and having a large contrast in the state of the two sides of the ball is the sine qua non of obtaining reverse swing, it will always be treated with suspicion, particularly when only one side in a particular series proves itself capable of reversing the ball.

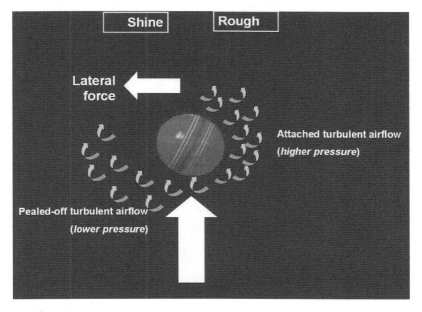

Reverse swing at high speed with an old ball. The angled seam causes the turbulent flow over the surface of the ball to peal off. The pressure drops on the shiny side relative to the rough side and the ball swings in the opposite direction to normal swing.

Why do not all bowlers reverse swing the ball? Even understanding the basics of *why* it happens, putting that into practice is not straightforward. Many bowlers are either not fast enough to obtain reverse swing, or unable to master the technique successfully. However, reversing the ball is now an essential weapon in the armoury of a Test team, particularly in conditions where conventional swing is difficult to obtain.

Before we finish this took at hydrodynamics, what about the other great mystery of bowling: the behaviour of the white ball, which we mentioned in Chapter 6?

The key words in Chapter 6 to describe the difference between the red and white ball were "the only difference in manufacture between the two balls is the dye used to colour the leather that covers the ball and the varnish that protects that colour". In the red ball, the natural oils of the leather are used to produce and to conserve the shine. In the white ball the colour is trapped in by a layer of varnish that helps to keep the ball clean and visible in flight. This varnish helps to conserve the shine on the new ball, slowing its deterioration and modifying its aerodynamic properties. As the ball roughens more slowly and conserves its shine for longer, quite naturally it will swing more and for longer than a conventional red ball. So, in fact, it is quite normal and natural that the white ball does behave somewhat differently. The change of the white ball half way through the innings as the ball gets dirty only serves to accentuate this different behaviour.

ENDNOTES

[1]   It is a remarkable indictment of Devon Malcolm's batting prowess that Phil Tufnell went in ahead of him. Tufnell and Malcolm are often regarded as being the worst numbers 10 and 11 ever to have played for England.

[2]   Opening partnerships of 28, 123, 40, 41, 168, 27, 39 and 29 for an average opening stand of 61.9 show that England were, if not flourishing, not in dire straits either against the new ball. Graeme Gooch averaged 48.0, Alec Stewart, who replaced the out of form Mike Atherton as opener in two Tests, averaged 56.7. Even Mike Atherton averaged 29.0.

[3]   This is how the BBC reported the developments at the time until it was finally confirmed, late in the evening, that Pakistan had forfeited the Test by not being ready to play when then umpires called play:
    http://news.bbc.co.uk/sport2/hi/cricket/england/5268250.stm

[4]   Bernouli, D.: 1738, Hydrodynamica

[5]   This theory of getting the ball to swing was extensively applied by the England seamers when they retained the Ashes in Australia in 1986/87. Even if the theory was hopelessly wrong, the bowlers got the result that they wanted.

[6]   This is sometimes called the "soldiers in a cabbage patch" model. If you have a row of soldiers marching across country across flat grass, they will all march at the same speed. Make that same row of soldiers enter a cabbage patch at an angle and the speed of march will reduce sharply. To maintain formation while the soldiers at one end of the line are still marching at constant speed, but those who enter the cabbage patch slow, the whole line must pivot on the end of the line closest to the cabbage patch and turn in towards it.

[7]   A case in point is the famous Starfighter. It had razor wings that were extremely thin, so that the difference in path and thus airflow velocity over the upper and lower wing surface is effectively almost zero, but still flew very effectively.

[8]   Why was Geoff Boycott, famed in the earlier part of his career for delivering big, banana swingers, a much easier proposition to play than his teammates John Snow or Bob Willis? The answer is, obviously, that Geoff Boycott's much slower deliveries gave the batsman far more time to adjust his stroke and gave him much less margin for error in his line and length. Even so, he proved to be surprisingly effective as a bowler for England in the 1979 World Cup, taking the same number of wickets as Richard Hadlee or Imran Khan

at a better average and strike rate!!! Remarkably, he had the sixth best strike rate in the competition of any bowler who bowled in more than one match.

[9]   Lyttleton, R.A.: 1957, Discovery, 18, 186-191

[10]   Mehta, R.D., Bentley, K., Proudlove, M., & Varty, P.: 1983, Nature, 303, 787-788.
He was working at Imperial College (London) at the time, but now works for NASA at Pasadena (California).

[11]   The Pakistan attack in that Test was Sarfraz Nawaz, Imran Khan, Asif Iqbal, Mushtaq Mohammed, Iqbal Qasim and Javed Miandad. With the top three being Majid Khan, Sadiq Mohammed and Zaheer Abbas and the gloves in the hands of Wasim Bari, this is arguably one of the strongest Pakistan sides in history, filled with legendary figures.

CHAPTER 9

IT'S THE APPLIANCE OF SCIENCE

"It's a photo finish and I don't have time to develop the photo.
NOT OUT!"

(Anonymous umpire of a First Class match in England, pre-television era,
judging a run out appeal)

Today, no umpire could get away with such a statement as the one quoted above, which genuinely happened in a match in England in pre-television days, when the umpire's authority on the field of play was still absolute and he could apply common-sense in rulings without being condemned for it. Now, the all-seeing eye of the television camera watches every ball, every move, every gesture on the field of play and scrutinises every decision. Even though technology has demonstrated that umpires get far more decisions right than they get wrong, no one wastes their breath on the correct decisions, but one bad one, however marginal it may be, can see an umpire vilified and especially if it sees a player's career ruined.

Much of the umpire's work is administrative. The umpire controls play; counts the balls in the over and decides when the ball is live and when it is dead; is responsible for seeing that the playing conditions for a match are followed, that the laws are upheld and that the spirit of the laws is obeyed. However, for a minimum of 270 deliveries each day the umpire must also maintain an intense concentration on the ball and the action and be prepared to make as many as thirty or forty instant decisions, based only on the evidence of his eyes and ears, supported only by his colleague at Square Leg.

Of course, umpires are only human and it is a natural human reaction to assume that any decision shown to be a mistake – often only after half a dozen slow-motion replays – is due to incompetence or, worse, deliberate bias. In any sport the umpire or referee is usually held as the scapegoat for misfortune and is a pretty easy target[1].

On an appeal from the fielding side, the umpire must make decisions of four basic kinds:

1. Line decisions – Was the batsman in or out of his ground when the wicket was broken?

2. Flight decisions – Did the ball carry on the full to the fielder and was the catch cleanly made?

3. Trajectory decisions – Would the ball have hit the wicket had it not impacted the batsman's body first?

4. Perturbation decisions – was the trajectory of the ball perturbed by a glancing blow against the bat, or against the hand holding the bat?

On occasion the umpire may have to decide on several points simultaneously: did the ball hit the edge of the bat and, if so, was the catch made cleanly? If the bat was not involved, would the ball have gone on to hit the stumps and thus, should the batsman be given out LBW? But, then again, if it is to be an LBW decision, did the ball pitch outside leg stump? If the batsman was hit outside off stump, was he playing a shot or not? We routinely request an umpire to make split-second decisions following a complex decision tree that would require a fair amount of computing power to calculate so and, unlike the critics watch at home, or from the commentary box, he gets just one chance to see the action and no slow motion replay so, "no pressure, Umpire".

LINE DECISIONS

Line decisions are probably the simplest for the umpire, although also possibly the most fraught with danger because technology can prove them right or wrong so easily, with no room for opinion. In the past the umpire would use the criterion that "the batsman gets the benefit of the doubt", hence the wonderful quote that opens this chapter. Until the 1980s it was felt, even by players, that provided that the decision was a question of 6 inches (15 cm) either way, the decision should go always in the batsman's favour. What does this mean from the umpire's perspective?

A fleet-footed batsman who is sprinting for the crease may reach a maximum speed of 25km/h, equivalent to 7 m/s. The human retina operates at a frame rate – i.e. renews the image – of approximately 16 frames per second (this is often termed "persistence of vision" and can vary from person to person, but this relatively slow renewal rate is the reason why we see continuous motion rather than jerking between frames when we watch a movie at 24 frames per second). In other words, between two "frames" of the umpire's vision, the sprinting batsman will move:

7/16 = 43 cm

In other words, in this case the umpire is interpolating to about one third of a frame on a reasonable doubt criterion. This makes this "6 inch" criterion totally reasonable: it is ridiculous to expect the umpire to be able to judge smaller distances than this at full speed without guessing.

Working in his favour, the umpire has the extraordinary instrument that is the human eye. The human retina has two types of cells that act as photodetectors of the image formed by the cornea and the lens: there are 120 million Rods (so-called because of their distinctive shape) and 6-7 million Cones, these last being responsible for our colour vision. The immense majority of the Cones, some 98%, are red and green sensitive and are packed in a tiny central region of the retina called the Fovea Centralis. The rest of the eye is packed with the Rods. The Rods, which give us our night vision and also sense motion, do not form a colour image, but are still strongly colour sensitive as they have no ability to detect red; at the same time, the small proportion of blue-sensitive Cones are also spread around this area of the retina, away from the Fovea Centralis.

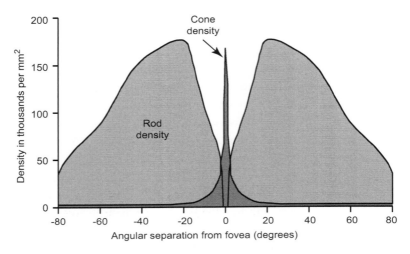

The density of Rods and Cones in the human eye. The central part of the eye – the Fovea Centralis – has excellent colour vision and resolution and also adapts rapidly to changing light levels, but has lower sensitivity to movement. In contrast, the Rods have good motion-sensitivity, but do not detect red; a cricket ball moving in the umpire's peripheral vision is strongly detected, but will appear much darker than the background of green grass to which colour the Rods do respond strongly[2].

As a result, the umpire's peripheral vision is very sensitive to movement, but totally insensitive to the colour red, so a red ball will simply appear to be dark grey as it comes into his field of view, projected with higher contrast against the bright green of the grass, which his eye will register efficiently. This motion sensitivity and inability to see red and green simultaneously in peripheral vision help the umpire to track the ball's progress and anticipate its arrival. Also on the umpires side is the human eye's sensitivity to lines. The human eye has an innate ability to resolve lines clearly. Why is this? The answer is that a line will cross many Rods and Cones on the retina, thus aiding its detection and resolution. The longer the line, the more the detectors on the retina that are crossed, meaning that the line's position is better determined by the brain: the long line marking the crease makes it easier for the umpire to make his decision correctly.

On the negative side, a further difficulty that the umpire faces is that he must move as fast as possible to get side-on to the crease as soon as a potential run-out becomes likely. If the umpire is not exactly side-on, foreshortening makes the uncertainty in his decision much greater: the further he is away from exactly side-on, the greater the uncertainty and the larger the likelihood that the decision will be incorrect.

Television samples the action at a higher frame rate than the human eye and, with the benefit of slow motion replays, it was possible to show that, on many occasions, an umpire who had given the batsman the benefit of the doubt was clearly wrong, if only by a small margin (almost always a few tens of centimetres). It also increasingly became the habit from the mid-1980s, even in Test matches, for close decisions to be repeated endlessly on giant screens inside the ground. As cricket became ever-bigger business, it became unacceptable for the umpire to give the benefit of the doubt while the television-viewing public and, increasingly, spectators at the ground watching giant-screens (and hence, the players themselves) were seeing, seconds later, that the batsman had actually been out of his ground. Even if the decision is very close, after four or five repeats in slow motion everyone in the ground and potentially millions of TV viewers will be cursing the umpire's stupidity, forgetting that he only sees the incident once and sees it both without warning and at normal speed.

Traditional television samples the image at 25Hz, about 50% faster than the human eye refreshes its image, so we see a smoothly moving image, as we do for the 24 frames per second that we view when we watch a film in the cinema. The television image refreshes at double this rate – 50Hz[3] – but the two images are interleaved, in other words, only half the image updates with each refresh. In video, the normal frame rate is 25Hz, without interleaving[4], giving a much better representation of movement and resolution in the vertical axis of the image. The resolution of the image though is not much better than the umpire's view, with about 10cm between frames for a sprinting batsman, about double the usual width of the line marking the crease; this line belongs to the fielding side, so the batsman is out if the bat is not past the line and into the crease. This limited resolution leads to occasions where the actual breaking of the wicket and crossing of the line by the bat may take place between frames, meaning that even an action replay is often not conclusive.

High-definition television uses double the normal frame rate, or 50Hz video output. This not only gives better resolution of movement, but also reduces the blurring of moving objects such as the ball, giving a much better definition of incidents. Even so, a ball moving at 150km/h out of the hand of a fast bowler will still move 80cm between video frames although, as almost all of the movement is directly away from the camera, the over the shoulder view of the ball means that, because of this foreshortening, the image of the ball is not as badly blurred as we would expect. Modern standard TV cameras can work at 72Hz and play back at 24Hz, to give a high resolution, high-quality slow motion image compared to normal video.

However, for the very fastest action and greatest clarity of image, innovations such as the SkyTV "Super SloMo" are desirable, allowing the ball to be followed with clarity down the pitch past the batsman. In the past, the BBC pioneered an experimental super-high frame rate of 300Hz for live sports broadcasts in High-Definition. It was felt that with increasingly large screens and high-definition, combined with fast action, even 72Hz is insufficient to show the action as clearly as would be desirable in many sports such as football and cricket where there is a ball in rapid motion. At 300Hz, even a sprinting batsman, running flat out and observed side-on, moves no more than 2cm between frames – a barely discernible amount. Sky's Super SloMo is the version of the BBC development most familiar to cricket fans, revealing as it does aspects of the game that are hidden at normal frame rates. The Super SloMo camera can transmit at such a high frame rate because it sends its signal over three data lines sequentially. This though requires a special, costly EVS recording system to take the three input signals and interleave them into a single picture stream for broadcast[5].

High velocity, high definition imaging also has a problem that makes it of limited practicality for extended use in real applications, as much greater frame rates than a few hundred Hz cause real technological problems quite apart from the elevated cost. A camera, filming a 2000x1000 image (approximately the size of the Sky Super SloMo frames) at 1000 frames per second, generates an impressive 2GB of data per second, requiring massive transmission bandwidth and data storage and retrieval capabilities. A single, six hour span of play, continuously videoed even with only two cameras, one at each end, requires a small matter of 86TB (TeraBytes) of data storage[6]. Sky uses an impressive five Super SloMo cameras: one "over the shoulder" and one at pitch level at each end, plus a fifth camera on the mid-wicket boundary.

Television viewers who have seen the Sky coverage of matches in England will testify to the impact of the super slow motion replays of particular incidents during the course of play. Watching a Super SloMo image allows you to discern the ripple of the batsman's muscles as the ball impacts them, the twist of the bat in the batsman's hand and the puff of dust as he hits the ball. Cricket becomes an art form at ultra slow speed.

These Super SloMo images are obtained by using a special camera developed by Sony, which generates 1080 frames per second on a 1920x1080 CCD detector. When 1080 frames per second are reproduced at the standard 25Hz, each second is stretched by a factor of 43, turning one and a half seconds of play into more than a full minute of images. Such a high frame rate allows the spin of the ball through the air to be followed: a thin edge, even one that is barely detectable to the stump microphone, will provide an off-axis impulse on the ball that will change both its rate and axis of rotation. If the spin of the ball is seen to change after passing the bat, you can be sure that there was an edge. If it does not, it almost certainly means that there was no contact with the bat (this we will examine further, below).

In principle there is no reason why even faster frame rates can be used. The very fastest cameras in existence can shoot several million frames per second, however, they suffer from the not inconsiderable problem that their extremely short exposure times require commensurately strong illumination, thus they can only be used for very special applications (to film a cricket match at one million frames per second would require a degree of illumination of the ground that would carbonise the players in short order, thus interfering with the normal progress of play).

After speculating briefly with the possibility of giving the on-field umpires a portable television to see the broadcast TV images of slow motion replays, it was decided to have a third umpire in a cabin, watching the TV feed. On a request from the on-field umpires[7], the Third Umpire will study the available images and indicate his decision. How and in what circumstances a review can be made is determined by the ICC regulations on the UDRS (Umpire Decision Review System or, more correctly, the "Third Umpire – Decision Review System" regulations) an eight-page document available on the ICC website at http://icc-cricket.yahoo.net/rules_and_regulations. php. The reliability of the decision depends ultimately on the quality of the TV images supplied by the host broadcaster of the match. In some cases, it may take more than a minute for a decision to be made, particularly for stumpings, which can be extremely difficult to call, even from TV images[8]. The criterion remains the same though: there must be clear evidence that the batsman was out of his ground; if, even after several replays, it is not possible to demonstrate that the dismissal was clean the Third Umpire must report that the evidence is inconclusive and the batsman should get the benefit of the doubt[9] (Section 3.3, paragraph (h) of the Third Umpire – Decision Review System regulations).

Most recently the ICC has authorised the extension of use of UDRS to on-demand reviews of decisions by the batsman or fielding side when the host broadcaster is able to supply the images to support its use and both sides are in agreement on its use. This allows the batsman or fielding side to request a review of a dubious decision using the available technology, in the same way that an on-field umpire may refer decisions where he is uncertain to the Third Umpire. Each side is allowed two unsuccessful reviews per innings: each time a review is made that upholds the umpire's decision, the side that has requested the review loses one of its two appeals; waste both and you are no longer able to challenge any decisions at all. This system has led to sides taking great care not to use their reviews frivolously and having to take a rapid decision what the chances are of a successful appeal against a decision, balancing it against the possible danger of losing a review that might be desperately needed later. This, in turn, has led to the very curious situation in one of the England v Bangladesh Tests in Spring 2010, of a batsman who must have had a good idea that he was really not out, NOT risking requesting a review although the television images clearly showed that the dismissal almost certainly would have been overturned on review!

Anyway, we are getting ahead of ourselves here. Line decisions are a non-issue now as, in reality, for them, technology had almost eliminated error, even before UDRS was trialled. Other types of decision though are very much still controversial. Let us go back to looking at how technology can help with other types of decision.

FLIGHT DECISIONS

Flight decisions, like line decisions, can be the easiest and the hardest for the umpire. Did that edge to the slips carry, or was it just short? Did the fielder's fingers slip underneath before it touched the grass, or moments later? Visually, no one has a better view of a slip catch than the Square Leg umpire who is close to the action and has an almost perfect side-on position, putting him in the perfect situation to judge the flight of the ball reliably.

To date, science has not managed to find a satisfactory solution to the problem of judging whether or not a catch has carried. Normally, for a slip catch, the umpire at the bowler's end will ask his colleague at Square Leg, who is viewing the action side-on and thus has a much better view of the flight of the ball to confirm whether or not a catch is clean, recognising the limitations of his own view.

One of the low points of 2010 Ashes series was three incidents of low catches that were claimed at critical junctures on the second and fourth days of the Second Test: two of them were given, one of them after the consultation with the Square Leg umpire, the other after referral to the Third Umpire. The third was not given after being referred to the Third Umpire. A look at the Internet blogosphere finds accusations that in one of these incidents Andrew Strauss was cheating by claiming a catch knowing that it had not been taken cleanly[10] and, certainly, Ricky Ponting conveyed clearly his feeling that Australia had been wronged on all three occasions. So, let us look at them and see where things went wrong and what, if anything, technology could do about it.

On Day 2 of the match, played at Lords, England started in a strong position at 364-6. Initially things went very well for Australia, with Andrew Strauss falling, playing no shot, to the second ball of the morning, followed in short order by Graeme Swann and Stuart Broad in the first three overs

Why are low, Slip catches much easier for the Square Leg umpire to call than for the umpire at the bowler's end? It is a question of perspective: the umpire at the bowler's end gets an extremely foreshortened view, almost along the line of the trajectory of the ball (top image); in contrast, the Square Leg umpire has an almost side-on view of the flight (bottom image). In each case the arrow shows the typical trajectory off the bat that the umpire will see; improvised game between a group of soldiers on a tour of duty, or Test match, it is clear that the Square Leg umpire is in a far better position to judge the exact flight of the ball and whether or not it has carried to the awaiting fielder. However, you can bet your last cent that, whether it be a Test match or improvised game, an Australian will play it hard and claim the catch if he thinks that he has a chance of getting the verdict from the umpire!

of play. From there on, things started to go horribly wrong for Australia as Graeme Onions and Jimmy Anderson, neither renowned batsmen, flayed the bemused Australian attack to set a tenth wicket partnership record against Australia at Lords and take England to an unexpectedly strong score of 425. Then, in his more usual guise as a new ball bowler, Jimmy Anderson added insult to injury in front of the approving gaze of the Queen (not known to be a cricket fan) and Prince Philip (who had been an enthusiastic cricketer himself in his younger days). First he removed Philip Hughes to an (uncontroversial) gloved catch down the leg side to leave Australia 4-1 and bring in Australian captain Ricky Ponting and then, more controversially, he claimed Ricky Ponting himself to make the score 10-2.

Here things get murky. Anderson bowled a fine, full-length, inswinging delivery into the batsman's pads, from where it flew to Andrew Strauss at First Slip. Strauss dived, rolled and caught the ball, inches above the turf. England appealed, initially it seemed for LBW. This appeal was rejected. With the appeal refused because the umpire ruled that there had been an inside edge onto the pad, England then, as was their right, claimed the catch. This was referred to the Third Umpire because the umpire at Square Leg was unsure that it had carried. After a delay, the Third Umpire confirmed that the ball had carried and the catch had been taken cleanly. Umpire Rudi Koertzen was vilified later for a bad decision, particularly in Australian media, but the incident illustrates just how difficult such decisions are both for the umpire and for technology to resolve.

Ponting was not best pleased to be given out. Why not? He felt that he had not hit the ball and the high-resolution slow motion image seemed to support his case. What is more, the Third Umpire could, *at the time*, only advise on whether or not the ball had carried and the catch was clean, not on the issue of whether or not it had hit the bat, despite the obvious doubts from the images.

Even after watching the incident several times at natural speed, in slow motion and in high-resolution slow motion it is not easy to say exactly what had happened. Tellingly, in the commentary box, former England captain, wicket-keeper and opening bat, Alec Stewart immediately called it as a fair catch exclaiming "to the naked eye that is out!" Only after seeing the high-resolution slow motion images (Super SloMo) could the same commentator say, with the benefit of 20-20 hindsight: "a poor decision from Rudi Koertzen". So, why did he say that? What in the slow motion images made him change his mind?[11]

The ball was full and angled into the pads. Height was not an issue and the only doubt was whether or not the batsman had hit it – in which case he could be out caught, but not LBW – and whether or not, if he had not hit it, the ball was sliding down leg side. Umpire Koertzen ruled out the lbw on the grounds that he ball had been hit and the TV images, although only in slow motion, confirmed that Andrew Strauss had taken the ball cleanly, so the Third Umpire, *who could only be consulted on whether or not the ball had carried*, confirmed that the catch was taken cleanly. What only became evident after the dismissal was confirmed, viewing the incident with the benefit of high-resolution slow motion, was that that when the bat twisted violently in Ricky Ponting's grip – normally a sign of a thick inside edge[12] – the movement and the sound were produced by the bat catching in the batsman's boot. By twisting in that way, the bat was taken out of the ball's path. However, it takes several repetitions of the incident to say that the ball *probably* missed the edge of the bat completely or, if it caught the edge, it was an extremely fine contact[13].

This incident was taken by some sections of the Australian media and fans to suggest that the Australian team were the victim of a string of consistently unfair decisions. The postscript that is rarely mentioned is the fact that Hawkeye – which we will meet later – showed clearly that if Ricky Ponting had not got an edge on the ball, he should have been given out LBW anyway because the ball was going on to hit leg stump. In other words, either way, he was out: only the mode of dismissal recorded in the scorebook was incorrect.

The problem became worse when, with England 130-2 later on in the evening, Ravi Bopara hit the ball to Nathan Hauritz at Mid On who took a wonderful tumbling catch: or did he? Bopara stayed at the crease because he was uncertain that the ball had carried (as is his right) and waited for a decision from the umpire. Once the batsman did not walk and without a clear view of the incident for either umpire, the umpire referred the catch and the television image, which was not edge-on and was looking slightly downwards, could not show clearly whether the ball had touched the turf or not before settling in the hands of the fielder.

The lack of clarity of television images in judging catches had already been shown to be a major issue in the Caribbean early in 2009 when umpire referrals were trialled in the West Indies v England series. It became obvious that if a disputed catch was referred to the Third Umpire, it would be impossible to tell whether or not it had been taken cleanly sufficiently conclusively to confirm a dismissal. The television images were never clear enough to confirm whether or not the ball had bounced before being taken by the fielder; similarly, the viewing angle would make even a clean low catch appear to have bounced before it reached the fielder. It became immediately obvious that any batsman who requested a review of a low catch, even if obviously clean to the naked-eye, would be given not out on referral. Batsmen would refer even catches that looked to be probably completely fair, in the knowledge that a reprieve was more than likely because the Third Umpire would find the television images introducing doubt where none should have existed. The Bopara-Hauritz incident was just one more case showing how inadequate television is for judging the validity of low catches.

However, the incident that really upset the Australians came on the fourth morning, with Australia chasing a small matter of 521 to win. Simon Katich had been caught to leave Australia 17-1 and in trouble[14]. This brought the captain in, once again, to bat with the increasingly beleaguered Phil Hughes, who was short on runs so far during the tour and had revealed a weakness against the short ball since receiving a brutal working-over from a fired-up Steve Harmison at Worcester in the match against the England Lions[15]. The two were just beginning to weather the storm when, at 34-1, Andrew Flintoff unleashed a thunderbolt that took the outside edge of Phil Hughes's bat and flew low to the England captain. The commentator exclaimed "that one is out! Beautiful catch from Andrew Strauss!" and the batsman started to walk. Then, mindful of the Bopara catch the previous day, the batsman hesitated and went back, waiting for a decision as England celebrated. The Square Leg umpire, Billy Doctrove confirmed to Rudi Koertzen that the catch was clean from his side-on perspective. It was only when the high-resolution Super SloMo showed the catch from its elevated front-on view (i.e. the one that the umpire at the bowler's end feels is inadequate to make a decision) that suddenly the commentators hesitated too because, from that foreshortened perspective, the catch was far from clear and, on referral, could not have been given[16] (yet, of course, we expect umpires to do better with just a single look at normal speed).

In all three decisions there is a common thread: to the naked eye the dismissal seemed quite clear. In none of the three cases did the commentator doubt the dismissal until after seeing the Super SloMo replays and in none of them is the evidence of the replays completely conclusive.

COULD TECHNOLOGY DO BETTER?

As it stands, technology has multiple limitations here:

- Insufficient and inadequate camera angles – the over the shoulder view is simply not suitable for judging if a catch has carried; a side-on view is necessary. This is exacerbated by the fact that the camera is often well above pitch level and looking down so as to get the best view without being blocked by the bowler and umpire, thus making it even harder to judge if the ball has carried or not. Similarly, there is no horizon to provide a reference for the ground level. To do better requires many more cameras with Super SloMo capability to be placed around the ground, so that the best view can be selected of any incident however, the cost of this would be totally prohibitive. Possibly a backward facing stump camera, situated near the base of the stumps to get a good ground-level perspective and a clear horizon reference, would be the best way to judge low slip catches, although it would be blinded when the wicket keeper is standing up.

- Insufficient resolution of image – on a normal TV screen the definition of the image is not high enough to judge whether or not the fielder's fingers have slid under the ball. Possibly this could be improved by high-definition television, although, in reality, this only offers about double the effective screen resolution to the viewer, whereas even more resolution is needed for total certainty as to whether or not the ball has carried, even if the camera angle is ideal.

- Insufficient speed of reaction – is it acceptable for play to be held up for several minutes each time there is a disputed catch, while the Third

Umpire tries to decide from multiple replays at different speeds and from different angles whether or not the ball has carried? What does that do to the authority of the on-field umpires? Many people, including at least some umpires, feel uneasy about this. In contrast, when a decision has to be made the broadcasters are placed in the spotlight and their images are made protagonist, adding to the drama, so they are not necessarily adverse to this enhanced role in the action.

Here we reach a "reductio ad absurdam" in which the desperate attempt to eliminate the possibility of error means that we try to eliminate the element of reasonable doubt. If after two or three replays one cannot be certain that a batsman is out, should he not always be given the benefit of the doubt?

The problem here is that the unscrupulous can create doubt where none should exist and, if the replays of the incident are in any way inadequate – and, in low catches, they almost invariably are – a batsman can exploit the system in his favour. No one likes seeing a batsman being given not out wrongly, early in his innings and then going on to make a big century. However, at least, if the benefit of the doubt is applied consistently, neither side should be able to complain of being unfairly treated. If technology is to be applied sensibly for low catches, the ICC must define to broadcasters the required image resolution and the required viewing angle(s) to be able to judge catches successfully and the broadcasters must find a way of supplying them.

This though leads on to a second issue: that of cost. The full suite of technology required to review decisions (Super SloMo, HotSpot, Snicko, Hawkeye, etc.) is not cheap to buy and operate and only the wealthier host broadcasters can afford it. If homogenous technology is required in all international matches in all countries, it may be necessary for the ICC to pay for it and then loan the complete set of equipment out to the host broadcasters as required. As frequently there are several simultaneous series being played, the costs to the ICC of buying, maintaining and transporting multiple sets of expensive cameras and analysis tools would mount rapidly. Whether or not Test and One Day International cricket will continue to be profitable enough to make commercially viable the huge outlays on sophisticated decision-reviewing equipment that would be required remains to be seen.

So, let us continue to examine in more detail the different types of decision required by umpires (apart from the issue of line decisions, which is now effectively resolved) and how can science and technology help to make the decisions more accurately.

TRAJECTORY DECISIONS

Would the ball have gone on to hit the stumps if it had not been intercepted by part of the batsman's body?

It sounds simple, but Leg Before Wicket (LBW) decisions raise more heat than possibly anything else in cricket because the umpire's decision tree is such a complex one. There are many elements to take into account. Just look at the chain of decisions that the laws require…

Law 36 states:

1. Out LBW
The striker is out LBW in the circumstances set out below.

(a) The bowler delivers a ball, not being a No ball

and (b) the ball, if it is not intercepted full pitch, pitches in line between wicket and wicket or on the off side of the striker's wicket

and (c) the ball not having previously touched his bat, the striker intercepts the ball, either full pitch or after pitching, with any part of his person

and (d) the point of impact, even if above the level of the bails
      either (i) is between wicket and wicket
      or (ii) is either between wicket and wicket or outside the line of the off stump, if the striker has made no genuine attempt to play the ball with his bat

and (e) but for the interception, the ball would have hit the wicket

Not only does the umpire has to judge the ball's trajectory, taking into account the bounce, swing, seam and turn, but also he has to judge the batsman's intentions when the impact is outside the line of off stump and be alert for an edge into the pads. He is also expected to fix his gaze on the bowler's front foot and then, without loss of concentration on the ball, transfer his gaze instantly down the pitch. What is more, there are more appeals for LBW than for any other form of dismissal, most of which are invariably turned down; if the ball hits the pad, the fielding side will almost always appeal, however optimistic that appeal may be. After each appeal the umpire must make the same careful calculation, based only on the evidence of his eyes and ears.

When the appeal comes, like a judge trying a criminal, the umpire must ask himself if the evidence is strong enough to secure a conviction. Just as there are tough judges and more lenient ones, there are umpires who are more willing to convict than others. Arguably one of the greatest umpires of all time was Harold "Dickie" Bird. Only a mediocre player at First Class level for Yorkshire and Leicestershire – in 93 First Class matches between 1956 and 1964, he averaged just 20.71, with only two centuries – before becoming an umpire, from 1973 to 1996. Dickie Bird eventually stood in 66 Tests and 69 One Day Internationals. Over his career he earned a reputation as an umpire who made few mistakes and who was, very decidedly, a "not-outer": in other words, he applied a very strong criterion of proof when upholding LBW appeals[17].

The fact that there is a strong element of opinion in giving LBW decisions, such that two internationals umpires can watch the same incident, with one giving the batsman out, the other not out, yet in the end both can feel satisfied of the fairness of their decision, gives a flavour of the minefield in which umpires operate here. Going back to the 1980s British astronomer and keen cricketer, Sir Bernard Lovell, retired Director of the famous Jodrell Bank Radio Telescope, was asked by the cricketing authorities to look at ways of helping the umpires with technology and particularly to judge LBWs. At the time, the television viewer only had television replays to help to judge decisions; although these help define the line of the ball, judging height and trajectory is more difficult from a single viewing angle, especially an over-the-shoulder view, although still sufficient to make many a commentator exclaim "that must have been close!" or "that looked like a pretty good shout" when a decision was turned down.

Although the Bernard Lovell initiative came to naught, it set people thinking of how technological aids to umpires could be made available. Eventually new ideas did appear. From 2001, a new type of replay became available to broadcasters and has been gradually extended such that in many Tests it is freely available to fans following a text description of play in the Internet via CricInfo[18]. This is the Hawkeye system. A variant of this Hawkeye system has now become a familiar and accepted part of watching tennis, trusted by players, officials and fans as a crucial element and arbiter of play. Hawkeye has also been proposed as the solution to phantom goal incidents in football where a goal is not conceded even though the ball has completely crossed the goal-line, or conceded when the ball has not crossed the line.

Hawkeye, in both its cricket and tennis versions, uses an array of television cameras connected to computers. A minimum of six cameras are required, surrounding the field of play so that the trajectory of the ball can be followed from a variety of different angles. Side-on views determine the flight of the ball in the vertical axis. Face-on determines its direction in the horizontal plane; the more lines of sight, the more accurate the final positional determination. Computer-modelling then calculates a full trajectory solution with an error of only a few millimetres. In the tennis version the only question that is asked of the system is where the ball landed with respect to the line. The cricket version is more complex and, apart from determining the point where the ball pitches, is required to include a predictive element of the future trajectory of the ball, which is the most controversial element of the system.

In the absence of seam and swing, a cricket ball's motion is essentially simple, damped harmonic motion whereby the height of the ball above the ground is determined by Newton's equations of motion, modified by the loss of energy of the ball on bouncing and the slowing caused by air resistance. In the absence of this damping, the vertical component of velocity would be

$$v_v = u_v + g\,t$$

Where

u_v = the initial vertical velocity component

t = the elapsed time

g = the gravitational acceleration (9.81 m/s^2)

In reality, the vertical velocity after pitching is determined by the coefficient of restitution of the ball, i.e.

$$C_R = -v/u$$

Where

v is the vertical velocity after pitching

u is the vertical velocity before pitching

The smaller the coefficient of restitution, the more speed that a ball loses on pitching. For a new cricket ball the typical range is $0.55 < C_R < 0.60$. As the ball gets soft C_R decreases and the ball becomes progressively slower and easier to play off the pitch.

As the vertical component of the velocity is usually relatively small compared to the horizontal component, the effects of air drag can be ignored here as a first approximation to the vertical motion. Similarly, the influence of gravitational acceleration turns out to be small – for a good length delivery from a fast bowler, the increase in speed due to the influence of gravity between release and pitching is less than 3m/s (about 10km/h).

In the absence of air drag – for example in a hypothetical future cricket match on the Moon – the ball's horizontal velocity would simply be constant until it pitches, at which point it would lose energy and thus velocity on impact with the pitch and then, assuming that the ball only bounces once, will continue to be constant until it hits the bat. In practice though, there is a strong deceleration on the ball from air resistance and thus the horizontal velocity is

$$v_h = u_h + a \times t$$

Where

    u_h is the initial horizontal velocity

    a is the acceleration due to air drag, in this case the acceleration will have
        be negative (i.e. the ball will slow)

    t is the elapsed time.

Finally, there is the lateral component of motion that may be caused by seaming after pitching, or swing from the moment that the ball leaves the bowlers hand.

As a first approximation the lateral velocity can be modelled as

$$v_l = u_l(t) + a_l\, t$$

Where

    a_l is the lateral acceleration (swing component)

    t is the elapsed time

and

    $u_l(t)$ is the lateral component of velocity imparted by the seam or the turn from spin after pitching. This velocity component is zero until the moment of pitching and thereafter may be assumed to be constant.

So, Hawkeye integrates the three measured velocity components to give the position of the ball as a function of time

$$S\,(x_t, y_t, z_t) \;=\; S\,(x_0, y_0, z_0) + \int_{0,t} (v_v, v_h, v_l)\, t$$

$S\,(x_t, y_t, z_t)$ is the x, y and z coordinates of the ball at elapsed time t

$S\,(x_0, y_0, z_0)$ are the initial coordinates of the ball at the time of release by the bowler and

v_v, v_h, v_l are the vertical, horizontal and lateral velocity components at elapsed time t.

Hawkeye takes a series of data points on the instantaneous position of the ball in each image frame and fits a series of equations of motion to them to model the ball's behaviour, including the second and higher order terms, as appropriate. As always, the final fit is only as good as the data that it is fitting and each solved component of the fit will have an associated error; the fit can never be perfect because the input data is never perfect. This is, to an extent, the reason why officials tend to distrust Hawkeye: the main reason why Hawkeye has not previously been offered as an umpire aid is this distrust of its *predictive element.*

When the trajectory of the ball is intercepted by the batsman's body, Hawkeye extrapolates the trajectory to the wicket. All it is doing is extrapolating the equations of motion into future time and thereby lies the rub. First, Hawkeye can only extrapolate the current movement and cannot anticipate how much the ball may have turned on pitching if the ball is intercepted on the full. Secondly, a small uncertainty in the position of the ball in the point of impact will become a much larger error the further that it is extrapolated into the future, particularly when the equations of motion have significant second or higher order terms[19], or the ball is only tracked for a very short distance before impact; if the batsman is a long way down the track when hit, the extrapolation is obviously far more uncertain, which is why umpires rarely give LBW when the batsman has come down the pitch.

In both cases Hawkeye can only assume that the observed trajectory behaviour will continue. If a batsman has taken a long step forward to a wickedly spinning delivery and is hit, full pitch, in front of the stumps, Hawkeye is unable to predict whether or not the ball would have turned sufficiently to miss the stumps on pitching as it has no future knowledge. Although Hawkeye does have the capability to calculate the error in the trajectory and thus the uncertainty in the point of impact on the stumps. in the classic television presentation this information is not shown. As a result, we should be very sceptical of any potential dismissal that just shows the ball grazing the stumps, or trimming the top of the bails.

What can be done to make Hawkeye information more understandable and useful?

There is one obvious innovation that could be implemented very simply. Given the known size of the ball, the known size of the stumps, the calculated trajectory of the ball and its calculated error, it would be a simple matter to calculate the probability that the ball will hit go on to the stumps. You then decide a criterion for dismissal and use it. Most people would say that if a decision is 50-50 it should go in the batsman's favour, so maybe an acceptable criterion is 67% probability – in other words, the ball is twice as likely to hit the stumps as to miss them – but it could be 75%, 80%, or even 90%; the important thing is to select a value and stick to it. In most cases there is little or no doubt that the ball will either hit the stumps, or miss them but, if the on-field umpire is uncertain, to get a whisper in his ear from the TV umpire over the radio that there is a 90% probability that the batsman is out would be a massive help in taking a decision and would not diminish his authority.

What has actually been done, where Hawkeye is being used for referrals, is to specify that when an umpire's Not Out decision is challenged, it can only be overturned if a 50% overlap criterion applies as detailed below (adapted from "Third Umpire – Decision Review System rules Section 3.3, paragraph (i), sub-paragraph (ii)").

- If an LBW appeal has been rejected and the batsman declared Not Out and the fielding side request that the decision be referred, the umpire's decision cannot be overturned unless Hawkeye shows at least 50% of the width of the ball hitting the stumps. Even if Hawkeye shows the ball grazing the stumps, that is insufficient to overturn the decision.
- If an LBW appeal has been granted and the batsman declared Out and the batsman requests that the decision be referred, the decision can only be reversed if no part of the ball is touching the stumps on the Hawkeye projection. If the ball is even just clipping the stumps, that is sufficient to uphold the umpire's verdict of "Out".

In other words, a sensible "reasonable doubt" criterion is applied such that unless Hawkeye is clearly in disagreement with the umpire, the umpire's decision stands.

PERTURBATION DECISIONS

Put bluntly, a perturbation decision is: did the ball hit the bat or not? Most perturbation decisions are simple – the ball hits the edge of the bat, there is a noise of leather on wood and the trajectory of the ball changes. No batsman is unaware that he has edged the ball unless the edge is extremely thin or there is double contact, for example, bat on ball and pad simultaneously, and, on most occasions, the edge is obvious to the umpire and fielders. More difficult and frequently controversial are the bat-pad decisions where there is a flurry or bat, pad, ball and fielder; there may be several noises and the umpire is encouraged to give the decision the way of the fielding side with a huge shout of "catch it", followed by a raucous appeal. At the same time the batsman will take great care to rub his arm or leg to indicate that the ball was nowhere near the bat[20]. Similarly difficult are the occasions when the ball passes very close to the edge of the bat, particularly on the inside as, visually, such fine edges may be almost impossible to detect with clarity, especially if the umpire hears several noises that may mask the telltale sound of the edge of the bat being brushed and thus has no clear audible guide.

However, in edged catches, many things happen that make technological aid simple to provide:

- Scattering – On impact with the edge of the bat, the ball will be deflected through an angle proportional to the distance between the point of contact and the centre of the ball.

- Frictional heating – Frictional forces from the grazing impact will heat the edge of the bat.

- Conversion of kinetic energy into acoustic energy – Part of the ball (and, if a stroke is played, the bat's) kinetic energy will be converted into acoustic energy (i.e. noise).

- Momentum transfer – there will be momentum transfer to the ball from the off-axis impact that will modify its rotational state.

SCATTERING

One of the bowler's chief methods of dismissal is to obtain an uncontrolled edged shot behind the wicket that carries to the wicket-keeper, or to the slip cordon. There are many ways of getting it: a batsman may play for non-existent spin, or the ball may spin more than he expects; the ball may seam off the pitch just enough to catch the edge of the bat; the batsman may misjudge the swing and play inside or outside the line; or the batsman may simply make a mistake and play completely down the wrong line. The ball can also hit the bat off centre with sufficient force that the rotational torque is greater than the frictional force exerted by the batman's grip, making the bat twist uncontrollably in his hand. Most normally, edges will be caught on the off-side in the arc between the wicket-keeper and the gulley or, in a very aggressive field, gullies. Inside edges will usually be deflected into the batsman's pads or, if very fine, more often than not will evade the wicket-keeper's despairing dive, frequently running for four (the so-called "Chinese Cut" shot).

The degree of deviation to get an uncontrolled edge is small. Law 5 states that the ball should be between 22.4 and 22.9cm in circumference. Law 6 states that the bat should be no more than 10.8cm wide at its widest. So, to hit the edge, the point of impact should be between half a bat-width and half a bat-width plus the radius of the ball, off-axis from the centre of the bat. If we take the average size, the radius of the ball is, of course

$$r_{ball} = 22.6/2\pi \text{ cm} = 3.6 \text{ cm}$$

So, if the batsman makes no correction for the movement of the ball, for a classic edge the bowler is aiming for a zone from 5.4cm to 9.0cm on either side of the centre of the bat[21]. This is the reason why you will often hear the commentator stating that "the ball did too much"; only a very small movement is required to take the edge of the bat. A big seam movement may look impressive, but all too often will evade both the bat and the stumps and thus be no real threat to the batsman.

The deflection of the ball is a similar to Rutherford scattering of alpha particles.

In 1909 Hans Geiger and Ernst Marsden had been experimenting firing alpha particles at thin metal foils. Most of the alpha particles were expected to pass straight through or, at most, be deflected through a tiny angle, some though were reflected almost straight back. Famously, Rutherford remarked

"It was quite the most incredible event that ever happened to me in my life. It was almost as incredible as if you had fired a 15-inch shell[22] at a piece of tissue paper and it came back and hit you."

From these results, Rutherford deduced that the atom had to have a tiny nucleus with a positive charge that repelled the alpha particles. Rutherford derived a formula for the scattering that described how many should be scattered through a given angle θ

$$N(\theta) = \frac{N_i nLZ^2 k^2 e^4}{4r^2 KE^2 \sin^4(\theta/2)}$$

N_i = number of incident alpha particles   n = atoms per unit volume in target
L = thickness of target   Z = atomic number of target
e = electron charge   k = Coulomb's constant
r = target-to-detector distance   KE = kinetic energy of alpha particle
θ = scattering angle

For alpha particles, substitute deliveries, for the atomic number of the target substitute the width of the bat, for the electron charge substitute the diameter of the ball and for the kinetic energy of the alpha particle substitute the speed and mass of the ball and you have a pretty good approximation to the scattering of cricket balls off a bat.

Of course, in the case of a ball hitting a bat, the "nucleus" is so large that many of the particles will be directed straight back at the bowler, whereas another large fraction will pass straight through to the detector (the wicket-keeper)! Where the scattering angle is large, or zero, the umpire requires no further aid, it is usually only in the case of a very small scattering angle where problems tend to occur and the aid of technology is required. So, let's look at some cases of what happens physically at small scattering angles.

FRICTIONAL HEATING

The thermal energy produced by friction is a simple function of the coefficient of friction and the normal force. In other words

$$E_{th} = \mu_k \int_x F_n(x)\ dx$$

Where

μ_k is the kinetic friction
F_n is the force applied in the normal direction
x is the direction of the ball's motion

The static coefficient of friction of leather on wood is quite low: 0.3-0.4[23]. Kinetic friction is almost invariably smaller than static friction and, for a fine edge, the force in the normal direction is small and applied over a very short distance (the thickness of the edge of the bat), thus the amount of thermal energy produced by friction is small. However, the thermal conductivity of wood is low, so the frictional heat produced is almost all radiated to the atmosphere from a small, concentrated region, rather than being dispersed by thermal conduction within the bat. For a soft wood, the thermal conductivity is $\approx 0.12\text{W m}^{-1}\text{ K}^{-1}$ (stainless steel, which is one of the poorer metals for thermal conductivity, is more than two orders of magnitude more thermally conductive than wood).

The heat energy will be emitted as blackbody radiation following Planck's Law:

$$u(\lambda,T) = \frac{\beta}{\lambda^5} \cdot \frac{1}{e^{hc/k_B T\lambda} - 1}$$

Where

λ is the wavelength
h is the Planck constant
k_B is the Boltzmann constant
c is the speed of light, and
β is a constant

The wavelength of greatest emission is given by the Wien Displacement Law:

$$\lambda_{max} = 2.898 \times 10^{-3}/T$$

Where T is the temperature in Kelvin.

The bat is at the same temperature as its surroundings and, whether we are in Durham in May, or Mumbai in November, the ambient temperature will be within about 5% of 290K, so it will radiate thermal energy peaking at close to 10 microns. This is the wavelength at which thermal imaging cameras work.

As we are at the peak of the black body spectrum, even a small temperature difference produces a large change in the intensity of emission. In other words, if we look at the edge of a bat that has just grazed the ball we will see that it is slightly brighter at 10 microns than the rest of the bat.

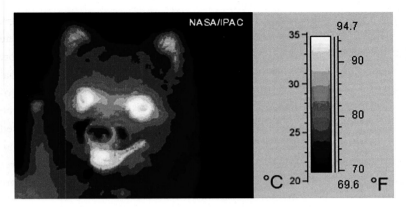

A thermal imaging camera image of a dog. Black and dark blue correspond to the weakest emission and thus the lowest temperatures. Yellow and white are highest emission and thus the hottest temperatures. The dog's fur is a good thermal insulator, so here the emission is lowest. In contrast, the eyes, the inside of the mouth and, to a lesser extent the ears, where there is no fur, show up as hottest. And yes, this dog has a healthy, cold, wet nose, as indicated by its darkness! HotSpot uses exactly the same technique to detect slight increases in the temperature of the edge of a bat heated by friction from a grazing impact with the ball. Image: NASA/IPAC.

This is the technique used by HotSpot. It is simply a sensitive thermal imaging camera capable of detecting the tiny increase in temperature produced by the friction of the ball against the edge of the bat. As the temperature differences to be detected are small and good resolution is needed, a particularly sensitive, large-format camera is required; this means that HotSpot technology is out of reach of many host broadcasters[24]. However, the presence of a temperature increase when the ball passes the edge of the bat is conclusive evidence of contact. Similarly, its absence, particularly when other techniques also fail to show evidence of contact with the bat, is as conclusive as it is possible to get that there was no contact between bat and ball[25].

CONVERSION OF KINETIC ENERGY INTO ACOUSTIC ENERGY

Not all of the kinetic energy lost in the contact with the bat will be converted into heat energy; part will be converted into acoustic energy. The impact will cause a shock wave that will set the air around the bat vibrating. This, the fielders and he umpire will detect as a characteristic, sharp, wooden sound. In contrast, if the ball hits the pad only, which is a yielding surface, the impact will last for longer and the sound produced will be more muffled. Unfortunately, things are not as simple as this. In practice, it can be extremely difficult for the human ear to distinguish between the sound of bat on ball, bat on boot, bat on ground and, even, a glancing blow of ball on stump (there are several cases in matches when a clear wooden sound has been heard, the close fielders have all gone up for a catch and, only after the appeal was turned down was it noticed, either from television images or from the characteristic red mark on a stump, that the ball had actually hit the stumps without dislodging a bail). Similarly, a sound that can be heard clearly by the wicket-keeper and slips, may not always carry to the umpire, especially if he is standing back to avoid impeding the bowler or there is some wind carrying the sound away from the umpire.

It was realised though that something that had been a rather controversial introduction to broadcasting some years earlier, namely the stump microphone[26], could help to determine whether or not there had been an edge. By analysing the feed from the stump microphone the type of sound: a sharp sound from bat on ball; a muffled sound from bat or ball on pad, could be identified reliably. When this is combined with Super SloMo, the point of impact can be identified with a high degree of confidence: If a sharp sound is detected as the ball passes the bat (see the example below), there has been an edge.

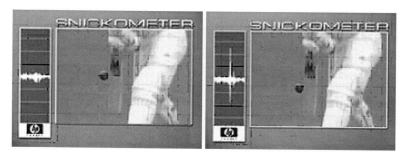

An example of how Snicko works. The feed from the highly sensitive stump microphone is displayed visually alongside the images. In the left hand image, just before the ball passes the bat, the trace from stump microphone is flat. As the ball passes the bat in the right hand image, we see a sharp acoustic signal typical of an impact between two solid objects. An impact with a yielding surface – for example, bat on pad – would give a more muffled sound of longer duration that would not necessarily coincide with the moment that the ball passed the bat.

Snicko is not one hundred percent effective – very fine edges may not be detectable but, in most cases, Snicko will be unambiguous. Its greatest inconvenience though is that the analysis of Snicko results is very slow because of the difficulty in aligning the stump microphone feed exactly in time with the television images. Results may take some minutes to come through, which is clearly unacceptable in a match situation and unfair to the players. This leads to Snicko being used *a posteriori* only to judge the correctness or otherwise of the umpire's decision rather than as a decision aid.

MOMENTUM TRANSFER

For the very finest edges, Super SloMo can eliminate almost all uncertainty. A grazing impact of the bat on the ball will transfer momentum to the ball in the form of an off-axis torque imparting angular momentum. This will modify the rate of spin of the ball. If the ball was not spinning before impact, the impact will impart spin. If the ball was spinning before hitting the edge, depending on the spin axis, the induced torque will either increase or slow the rate of spin and may change the axis of rotation.

Rotation Axis of Ball

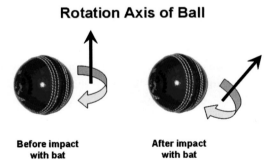

Before impact
with bat

After impact
with bat

A simple example of how angular momentum transfer will affect the spin of the ball on contact with the edge of the bat. Suppose that the spin axis is vertical before impact (left), contact with the edge of an inclined bat will impart angular momentum that will sum vectorially and thus tilt the spin axis (right frame).

If, in the Super SloMo replay we see a significant change in the direction or rate of spin of the ball on passing the outside edge of the bat, even if the signal from Snicko is inconclusive, it is conclusive proof that the ball has touched the edge of the bat. However, the inverse is not necessarily always true: it is *possible* in some circumstances, for the ball to touch the edge of the bat and not produce an obvious change in the rotation of the ball.

OBTAINING A CONVICTION

In conclusion, even if there is no detectable deviation of the ball, if we detect a sharp sound as the ball passes the bat, see the spin of the ball change and see a warm spot on the edge of the bat closest to the ball, we can say with a high degree of confidence that the batsman did hit the ball. However, just one positive result, unsupported, may not be and probably should not be enough to convict a batsman of dangerous driving.

The technology exists to aid the umpire in taking decisions. This technology is based on sound physical principles and, when used sensibly, can be of great assistance to the umpire. Where officials and administrators are reluctant to avail themselves of the technology the usual excuse is to cite doubts about its reliability. These doubts though often mask a more general concern about the trial by television of match officials – in some cases this has harassed officials out of the game – and diminution of their authority by technology. There is also a well-founded concern that constant referral of decisions will slow the pace of the game too far. One possible option to mitigate the former would be for the match officials to have access to enhanced technological assistance (additional information and cameras angles) above that which is broadcast publically, so that they can feel more in control and less on trial. Similarly, greater consultation between on and off-field match officials, with the television umpire able to offer discrete advise at the request of the umpire, may also help. More fluid communication would help to speed up decisions, particularly where the umpire just wants confirmation that his initial impression of an incident is correct[27]. In this, the rugby solution has a lot to commend it: the match official simply asks the television official for confirmation of his impression of an incident "is there any reason why I should NOT award a try?"

Endnotes

1   Geoff Boycott once said, very perceptively, that batsmen are quick to complain when they get a rough decision, but are never heard moaning "that was a dreadful decision, I was clearly out!" (although one or two batsmen have been known to come clean in the dressing room afterwards).

2   You can test this quite easily. Pick a reasonably large and strongly red object and then look well to one side of it so that it is in your peripheral vision. At some point you should notice that the red colour is no longer evident, even though you can still see the object clearly.

3   In the USA, the frame rate is 60Hz.

4   Technically, this reproduction rate is called 25p rather than 25Hz, demonstrating that it follows the international standard. International standard 75Hz, without interleaving, is 75p, etc.

5   You can find more detail about the technology at http://www.evs.tv/

6   It is a measure of how far technology has advanced that we can even consider such applications. When the BBC first introduced the Action Replay, as it was dubbed by the commentators, the storage was such that a decision had to be made by the Director every thirty seconds as to whether to use the previous thirty seconds of action before it was erased.

7   Indicated, logically enough, by drawing a TV screen in the air.

8   In stumpings, unless the batsman is clearly out of his ground, the TV umpire must judge too whether the part of the batsman's foot that was over the line was grounded, or not. This may sometimes be impossible to decide clearly without high-definition, slow motion images.

9   One of the problems in the initial trial in the 2009 Caribbean series between the West Indies and England was that there was not clear criterion for overruling a decision leading, in some cases, to worse miscarriages of justice than had the referral system not been present, with a probably correct decision being replaced by what appeared to be a probably incorrect one.

10 The accusation of cheating, in this case, means claiming a catch in the full knowledge that the ball had bounced first and thus deceiving the umpire and obtaining an inexistent dismissal to gain an advantage for his side. Strong stuff.

Not everyone will be convinced that the fielder deliberately tried to mislead the umpire. If the fieldsman's fingers have been jammed against the ground by the ball he will most certainly know that the catch was clean. In contrast, it is very difficult for a fielder to tell from his foreshortened perspective, whether the ball has been taken on the full or, in contrast, has bounced just in front of him. There again, even if the fielder suspected that the catch was not clean and has not passed on this information to the umpires, some would argue that it is no worse than a bowler constantly appealing for LBW knowing that the batsman was almost certainly not out, or a fielder appealing constantly for bat-pad catches to pressurise the umpire into error? When everyone "cheats" a little systematically no one can claim to be clean in this respect.

11   The fact that the commentator called it "a poor decision" was, in itself, unfair having only moments earlier called the catch clean and fair. The viewer will remember the criticism, not the original comment.

12   Later the same day, when England batted again, Kevin Pietersen survived a close lbw shout through a thick inside edge where, to the TV viewer, the telling evidence was the bat twisting violently in his hands due to the torque applied by the off-axis impact of the ball.

13   After watching time and time again, I am still not absolutely certain that the ball missed the bat. On some replays I convince myself that it did shave the bat. On others I am convinced that it missed. The umpire gets one look at full speed: how is the umpire expected to do better at judging what happened? However, in all the discussion of whether or not there was an edge onto pad, one critical aspect is forgotten… it was only that perceived edge that stopped Rudi Koertzen from giving Ricky Ponting out lbw!! Maybe the batsman was given out to the wrong appeal, but he should almost certainly have been given out one way or the other. However, this eventuality is now contemplated in Section 3.3, paragraph (f) in the Third Umpire – Decision Review System regulations, specifically tasking the Third Umpire with informing the on-field umpire if the mode of dismissal is not the correct one.

14   The fact that the replays showed that the delivery from Andrew Flintoff had been a no ball that was not spotted by the umpire did little to help their humour and sense of injustice. Now, on referral of a decision, the first check is of whether or not the delivery was a no ball (Section 3.3, paragraph (g) in the Third Umpire – Decision Review System regulations). Furthermore, an on-field umpire may request that the Third Umpire check for a No

Ball in case of an apparently fair dismissal, should he have any lingering doubt about the legitimacy of the wicket-taking delivery.

15 Some experts have argued that it was Steve Harmison's bowling at Worcester that won the Ashes for England. Phil Hughes had scored a small matter of 415 runs in his first three Tests against South Africa, in South Africa, the previous winter (160 runs more than the next highest Australian batting aggregate in the series), with two centuries and a fifty in six innings, before causing absolute mayhem in his short stay for Middlesex. There was a lot of criticism both of Middlesex's decision to offer him a short contract and thus acclimatise himself to English conditions and to the ECB for permitting it, with pundits fearing that he would continue to score huge numbers of runs through the Test series. However, the Phil Hughes bandwagon hit the buffers, first when Steve Harmison subjected him to a tremendous bombardment of short-pitched bowling, dismissing him for 7 and 8 in a match where Australia scored 796 runs for the loss of just 14 wickets and then when the England bowlers, and particularly Andrew Flintoff, made him look very uneasy against the short-pitched ball in the first two Tests. Although Shane Watson scored heavily as a stand-in opener when Hughes was dropped after the 2nd Test, he did so through consistent scores rather than big innings (none of Shane Watson's scores were among the twenty-five highest scores made by batsmen in the series). The fact that Australia had been forced to alter their plans was a huge moral victory for England, showing the Australians to be vulnerable to aggression, quite apart from removing a player who had shown himself capable of producing large, match-winning innings.

16 Ricky Ponting's complaint was: "why was the Bopara catch referred, but the Strauss catch was not?" The answer is that in the former, the Square Leg umpire could not say that he had had a clear view of the catch and could not then confirm that it was clean, while in the latter he could. While it is easy to feel some sympathy with Ricky Ponting's view that the referral of catches should be done consistently, he had objected strenuously to the referral of the Bopara catch and thus was equally guilty of an inconsistent approach to the issue. Here, there is no moral high ground for either side.

17 Curiously, for someone who gave so few LBWs, Dickie Bird's very last decision in Tests was to give England wicket-keeper Jack Russell out LBW to Test debutant Sourav Ganguly in the 2nd Test at Lords in 1996. It was said that when Dickie Bird gave an LBW decision often he smiled as he raised the finger: the smile of the cruel executioner? No! It was "the smile of a man who has seen the truth", someone who knew that the decision that he has

just given was correct (this is a beautiful description of an umpire who had the respect of the players).

18   http://www.cricinfo.com/

19   This is a very well known phenomenon. Try using a polynomial fit to a set of data points and then extrapolate that fit beyond the data. However well the polynomial fits your data, its predictive power will rapidly become zero as soon as it is beyond the fitted part of the curve delimited by data.

20   This type of activity led to one frustrated umpire of recent vintage, when asked his opinion of a side, to respond "they are all cheats, nice cheats, but cheats."

21   In reality, the target zone is larger because, if the ball hits the bat off-axis it will tend to twist in the batsman's grip, thus inducing an uncontrolled edge even if the impact of the ball is closer to the centre of the bat. The design of the bat is now a critical factor in controlling the scattering of the ball off the edge of the bat.

22   15 inches = 37.5 cm: the size of the largest naval artillery shells in the First World War.

23   The Engineer's Handbook.

24   For the Bangladesh home series against England in winter 2009/10, the host broadcaster was reputedly quoted a price of 10 million dollars to hire a HotSpot system of two cameras for the duration of the series. Unsurprisingly, the system was not used.

25   Unfortunately, various incidents have shown that the combination of HotSpot and Snicko is not necessarily infallible. If both show evidence of contact with the ball, or both fail to show it, this can be accepted to be conclusive. For some very fine edges though only one or the other may register the contact, leading to an ambiguous result, in which case tradition suggests that the batsman should receive the benefit of the doubt.

26   There were fears that some of the language picked up by the microphone could be too ripe for broadcast and, at least initially, there was a sound engineer listening to the feed from the microphone with his finger on a delay switch to cut the broadcast, if necessary.

27   The possibilities were brilliantly illustrated by Aleem Dar in the 4th Ashes Test v Australia at Melbourne (December 27th 2010). Matt Prior of England edged the ball to the wicket-keeper, but Aleem Dar suspected that bowler Mitch Johnson had overstepped. The umpire himself referred the decision to the TV umpire, Marais Erasmus, who duly confirmed that

his suspicions were correct. The decision to give the batsman out was reversed and Aleem Dar only gained even greater credibility in the eyes of all concerned.

Chapter 10

Putting some spin on it

Q: Are you a tortured genius?

A: Only when they sweep my arm ball!

Peter Roebuck asks Vic Marks what it is like to be a spin bowler[1]

Seeing a great fast bowler in flight is one of the great spectacles of cricket, but seeing a master spin his web, leaving fine batsmen groping as if blind for a ball that rarely seems to be where it was expected to be, is possibly even more compelling. I grew up watching the elegance and guile of Bishan Bedi, Derek Underwood and Intikab Alam plying their trade. Be-turbaned Bedi is remembered by my generation as one of the most beautiful bowlers to watch and an extremely popular player, who was a fine servant both to India and to the county that he represented, for many seasons [Northamptonshire].

Englishmen of my generation though, are scarred by names such as "Shane Warne", "Abdul Qadir" and "Muttiah Muralitharan" whose long Test careers seem to have consisted mainly of torturing hapless English batsmen into surrendering their wickets while the scoreboard operator sat bored, seeing the score change only when a wicket fell, or the bowler at the other end conceded a run. The appalling pain was masked by a horrified fascination and grudging admiration. Winning the Ashes in 2005 was so much more fulfilling because Shane Warne was still in the Australian side and just as good as ever. With a fast bowler you might get a loose delivery, or edge one to the boundary to keep the scoreboard ticking over, but what do you do when Muttiah Muralitharan has figures of 31.4-15-46-7 in the first innings and then, shamed by his profligacy, manages to be more economical in the second innings, with barely any reduction in his wicket-taking threat[2]?

Spin bowling is also a speciality that is more tolerant of deviations from the norm of sleek, lithe, well-muscled forms than fast bowling. Spin bowlers are allowed to have the odd extra kilo around the middle and to continue when most other players have been retired for a few years. No one was surprised when an Australian teammate made the comment that Shane Warne thought that a balanced diet was "a hamburger in each hand". People were simply intrigued when Bermuda's Dwayne Leverock registered on the Richter Scale as he came in to bowl in the 2007 World Cup. And it was par for the course when Shaun Udal stunned India at Mumbai in 2006 with a match-winning performance at the tender age of 37, having been unexpectedly plucked out of county cricket in the absence of Ashley Giles through injury[3], or when Ray Illingworth came back to captain Yorkshire when into his fifties. Spinners are expected to be non-conformist.

Back in Chapters 4 and 5 we had a look at how the ball behaves when bowled. For a fast bowler you can make a quite reasonable approximation that the gravitational deflection is small. It is even a fairly good approximation when a spinner fires the ball in with no attempt to flight the ball. What happens though when you have a genuinely slow bowler in action? During the Border-Gavaskar Trophy in India in 2010 the Australian off-spinner Nathan Hauritz was delivering some balls as slowly as 72km/h (45mph) and pitching the ball in a narrow range around 4 metres in front of the stumps. Of the 660 deliveries by the Indian spinners in the Australian first innings, less than 3% bounced over stump height and most of those only marginally, which goes some way to explaining why Allan Border was so surprised at Edgbaston in 1981 to find a delivery from John Emburey climbing at his face (see Chapter 5). Slower and shorter deliveries were actually falling again after bouncing, before reaching the stumps rather than climbing up at the batsman. It is with spinners that we see that the ball's trajectory genuinely is damped harmonic motion with a parabolic path and, instead of delivering with a downwards deflection, the spinner frequently releases the ball at an angle above the vertical as we can see from a representation of Nathan Hauritz's slower ball (opposite, top).

Let us now look at a more exact representation of a fast bowler's trajectory through the air (opposite, bottom) than the one presented in Chapter 5. Here we take into account both the slowing of the ball due to air resistance and gravitational deflection.

In fact, as you can see, for a good length delivery at 90mph (145km/h), the path of the ball is so close to a straight line that the approximation that it is one

is a very good one indeed. For a good, high action the ball deflects down at 10°
from the bowler's hand and deviates very little from a straight line trajectory.

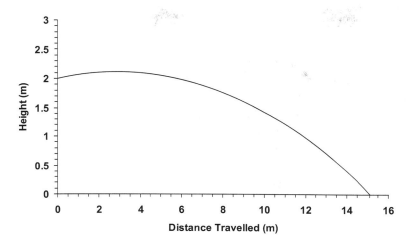

A representation of the trajectory of a 45mph (72km/h) Nathan Hauritz delivery taking into
account air resistance and gravity. The ball rises for nearly 3 metres after release, before
dropping to pitch 4 metres in front of the stumps.

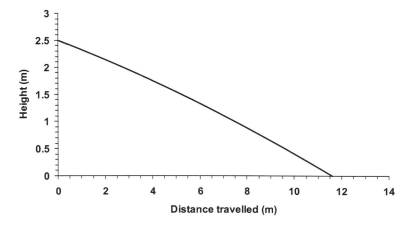

The trajectory of a ball bowled from a height of 2.5 metres by a fast bowler with a high action,
dropping the ball on a good length. Even when including the effects of air resistance, the trajectory
scarcely deviates from a straight line before pitching because the gravitational acceleration has
little time to work. The downwards deflection at release is 10° from the horizontal.

Some spin bowlers are surprisingly rapid. Derek Underwood, the England spinner of the 1960s and '70s was often described as left arm medium pace, being effectively unclassifiable in standard spin-bowling terms. One of the fastest spin bowlers playing today is Pakistani Shahid Afridi. Afridi clocks up some surprising figures on the speed gun. During the 1st Test v Australia at Lords in 2010, his fastest deliveries were clocked at 77.2mph (124km/h) by Hawkeye and he was regularly releasing the ball at over 60mph (97km/h). As he is quicker than most spin bowlers, Afridi tends to bowl a shorter length, pitching the ball about 1 metre further away from the batsman. His fast ball is released at a slightly shallower angle of −6.5°, but its flight is still well approximated by a straight line.

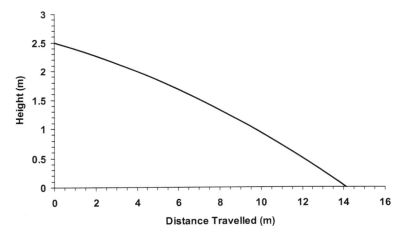

The trajectory of Shahid Afridi's fast delivery following a high delivery action. Although there is a more obvious deviation due to gravitational acceleration, the trajectory of the ball through the air is still reasonably well approximated by a straight line. Afridi tends to pitch a little shorter than other spinners, requiring a downward deflection of 6.5° on release.

Not all spin bowlers are as quick as Shahid Afridi, or bowl with such a high action. Many spinners effectively bowl from a crouch, with a much lower release point to allow them to give the ball some air. Cricket fans will remember the extraordinary "frog in a blender" bowling action of South Africa's Paul Adams. His release came from a very low height, as he was face down at delivery with his head close to the pitch. This meant that he had to flight the ball much more. Suppose he delivered a slower ball at 40mph (64km/h), what trajectory would it take? The answer is that, to pitch on a good length, when released from 1.2 metres above the pitch, the ball has to be released upwards at a 10° angle of deflection and will rise half a metre into the air before reaching the apex of its trajectory and starting to fall, 5.5 metres after release. Such a trajectory adds a new problem for the batsman, of estimating the ball's flight through the air when very nearly at, or even above eye level, something that is innately complicated for the brain to do and which has led to cases of batsmen being bowled ducking under slow beamers[4]. The ball will also take 0.9 seconds to bounce after delivery: an age for a batsman whose instincts are to attack the ball, possibly causing him to overbalance as he waits and thus be vulnerable to being stumped. It also causes the ball to fall at a much steeper angle to the vertical before pitching. This combination of factors gave batsmen a series of unfamiliar problems to deal with, quite apart from the distraction of the delivery action.

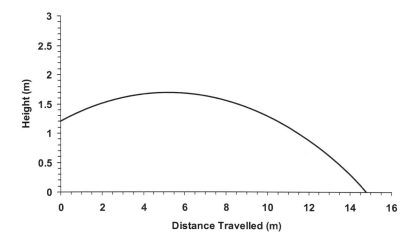

How to bemuse a batsman. The trajectory of a slow, flighted delivery from Paul Adams. When first introduced to international cricket Paul Adams of South Africa was extremely effective, taking 31 wickets at 25.1 in his first seven Tests, despite five of those Tests being against India: the best players of spin in the world. His unusual delivery action and very low delivery point posed considerable problems to batsmen initially, with the ball close to the batsman's eye line for much of its course. However, it required great accuracy and, once the surprise element was lost, was much less effective. After taking just 18 wickets at 52.3 in his last eight Tests, spread over five series in 2003 and 2004, he never played Test cricket again and even struggled to hold a place in his provincial side[5].

Even more bemusing results can be obtained by tossing a ball into the air very slowly – the so-called "donkey drop" – which can only be bowled successfully if the ball is released from around or below shoulder height. The ball stays above the eye level of the batsman until only a few metres before pitching and then drops nearly vertically. This was the bowling technique of Norman Teer, founder, patron and captain of the Mendip Acorns cricket team, whose exploits were immortalised by Peter Roebuck in the chapter entitled "Something Fell from Heaven" of his book "Slices of Cricket".

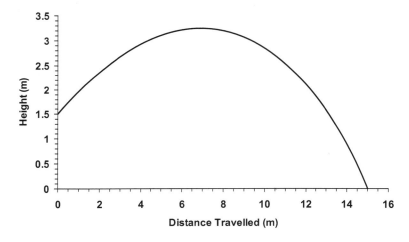

An example of a Norman Teer "donkey drop", achieved by bowling the ball very slowly from around or below the shoulder on a steeply upward path on release, such that the ball is above the eye line of the batsman for nearly all its flight and falls almost vertically. This makes judging its length extremely difficult. In this example the ball is released at 31mph (50km/h) at an angle of 25° above the horizontal, from about shoulder height and takes a remarkable 1.4 seconds from release to pitch. This bowling technique is similar to the underarm "lob" bowling that was still used occasionally, even in Test matches, at the start of the 20th Century, although lob bowlers would often use even more extreme trajectories, attempting to gain as much altitude as possible.

Peter Roebuck commented that there was nothing innately funny about Derek Underwood and Norman Teer bowling in tandem, as they did on one Mendip Acorns tour of the Caribbean, but spectators would come from miles around to watch the matches and would literally roll with mirth at the contrasting bowling styles of the two. This type of bowling harks back to the so-called "lob" bowling that was common in the late 19th Century. One of the most famous occasions that it was used in Tests was the Oval Test of 1884. On the second afternoon, with declarations not yet allowed under the laws and Australia building up a massive total, the Hon Arthur Lyttelton, the England wicket-keeper, who had already bowled overarm medium pace briefly and erratically the previous day without success, took the ball and produced a devastating spell of 4-8 with slow, underarm lobs, despite keeping his pads on to bowl. Australia collapsed from 532-6 to 551 all out, but the match ended in a draw[6].

A lob bowler would often attempt to launch the ball to a great height and get it to pitch BEHIND the batsman, dropping straight onto the top of the stumps. This required remarkable accuracy but, if the bowler got the trajectory right, it would be almost impossible for the batsman to avoid dismissal. The exploits of lob bowlers such as Digby Jepson "The Lobster", who would deliver the ball from a crouch, with his hand only inches above the turf inspired a wonderful short story by Sir Arthur Conan Doyle called "The Story Of Spedegue's Dropper" about a young lob bowler who became the saviour of the England cricket team[7]. Although overarm bowling had only been legalised in 1864 it took over from conventional underarm and lob bowling quickly; the last time that a specialist underarm lob bowler was selected for a Test was George Simpson-Hayward for

An example of lob bowling. Walter Humphreys photographed in the act of delivery of a lob [almost certainly a posed photograph]. Note the unusual grip of the ball. This photograph was taken around 1897, when lob bowling was still quite common in cricket at all levels. From K. S. Ranjitsinhji, "The Jubilee Book of Cricket, Third Edition" (from a photo by E. Hawkins & Co., Brighton).

the 5th Test in South Africa, which started on March 1st 1910. He took 23 wickets at 18.3 in the five Tests of the series with his viciously spun off-spin. Now though, lob bowling as practiced by Conan Doyle's fictitious Spedegue would not be permitted, as underarm bowling has been outlawed and any delivery from a slow bowler that passes the batsman on the full above shoulder height must now be called a No Ball for height under Law 42.6b(ii).

Flight is not the only way that a spinner will manipulate the trajectory of the ball through the air. Lateral drift through the air is also an important weapon. Anyone who has ever thrown a table tennis ball knows that it is almost impossible to get it to fly straight: it will almost invariably veer off laterally. This is because however carefully the ball is thrown, there is almost always some degree of spin imparted to the ball when it leaves the thrower's hand. For a right-handed thrower the natural effect is for the ball to have some anti-clockwise spin when viewed from above, meaning that, from the thrower's point of view, the right hand side of the ball is moving through the air faster than the left hand side because on the right the spin *adds* to the velocity of the ball and on the left it *subtracts*.

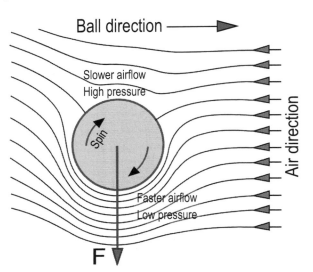

The Magnus Effect on a spinning ball. The ball's spin couples with the air around it, accelerating it in the direction of motion. As faster airflow corresponds to a lower pressure, a force (F) is generated towards the side of lower pressure that is spinning in the direction of spin. Thus the Magnus Effect allows an off spinner to drift the ball away from a right-handed batsman, while a leg spinner will drift the ball in to a right-hander. Adapted from an image prepared by Bartosz Kosiorek (Wiki Commons).

As air has viscosity, the surface of the ball couples with the air flowing around it and creates a faster airflow over the side of the ball that is moving with the spin. This is the Magnus Effect, first described by Sir Isaac Newton in 1672 while observing the effect of spin on Real Tennis balls[8], but popularised by the German physicist Heinrich Magnus in 1852. The faster airflow creates a low pressure area over the spun-up surface of the ball, leading to a lateral force from the high pressure side to the low pressure side. The ball will tend to swing in the air in the opposite sense to the spin.

The practical consequence of the Magnus Effect is that an off-spinner will tend to drift the ball away from a right-handed batsman, while a leg-spinner will drift it into a right-handed batsman. However, by changing the wrist position to apply top or backspin, a bowler can also induce the ball to hover or to dip, changing the length of the delivery unexpectedly. This was one of the abilities attributed to Tony Greig when he switched from seam bowling to off-spin with extraordinary success during the England tour of the Caribbean in 1974[9].

The strength of the Magnus Effect on a ball can be calculated by

$$F = S \, (\omega \times v)$$

Where ω is the angular velocity vector of the spin, v is the velocity vector of the ball, $\times$ denotes the vector cross-product and S is the average coefficient of resistance of air across the surface of the ball.

The Magnus Effect is a phenomenon that we have seen hundreds and probably thousands of times without realising. The same effect that causes a golf ball to hook or slice because an off-centre contact has put heavy side spin on it, will allow a tennis ball to appear to float slowly over the net when hit with heavy backspin. Any table tennis player knows that a ball returned with heavy chop (i.e. backspin) will tend to fly long off the end of the table while, in contrast, a ball hit with heavy topspin will dip strongly and will curve into the table when all logic says that it should fly long. And, of course, though not famous as a physicist, David Beckham's knowledge and application of the Magnus Effect has even inspired a film title[10].

We have now followed the ball out of the bowler's hand, through the air to the pitch. What happens when it hits the pitch?

Obviously, the bowler would like the ball to turn: that is, to deviate from the extrapolation of its trajectory at the moment of pitching. Whether it does or not and how much will depend on numerous factors.

- The velocity vector of the ball when it pitches.
- The angular velocity vector of the ball's spin.
- The coefficient of friction and duration of contact between ball and pitch.

A ball will not spin unless it grips the surface because unless it does it cannot convert its spin angular momentum into velocity. Similarly, the degree of turn depends on a vector sum between the velocity of the ball and the angular momentum so, the faster the ball is bowled, the smaller the influence of the spin on the final direction of the ball.

The angular velocity vector depends on the degree of spin imparted on the ball by the bowler and on the orientation of the axis of rotation. If we define that the x-axis is along the pitch in the direction from bowler to batsman, the y-axis is across the pitch and the z-axis to be the vertical. If the ball spins around the y-axis we will get topspin or backspin. If the ball spins around either the x or z-axis we will get side spin.

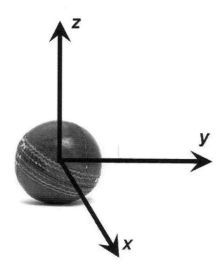

The three possible spin axes for a cricket ball. The x-axis is along the pitch. The y-axis is across the pitch. And the z-axis is the vertical. Rotation around the x and z-axis gives side spin, while rotation around the y-axis gives top (or back) spin.

If a ball is delivered with topspin, various things will happen. We have already seen that the Magnus Effect will make the ball dip through the air. Then, on hitting the pitch, spin angular momentum will be added to the velocity vector of the ball. The ball will speed up and keep low. From the batsman's point of view, the ball will pitch further away from him than he is expecting and then will hurry off the pitch and not bounce as high as he expects. The bowler thus has three opportunities to defeat the batsman's stroke: beating him through the air so that he does not reach the pitch of the ball; making him late on his stroke because the ball is faster onto him than expected; or slipping under the bat if he plays for non-existent bounce. The perfect top-spinner (from the bowler's point of view) will catch the batsman not far enough forward, scuttle through low and scud into the pad, trapping the batsman in front.

If the ball is delivered with backspin, it will tend to float and pitch closer to the batsman than he is expecting. The spin angular momentum will subtract from the velocity vector and the ball will stop on the batsman; it will also grip and climb, bouncing much higher than expected. Again, the bowler has various opportunities to bamboozle the batsman, particularly if he can be persuaded to go through with his shot too early, or be surprised by the height of the bounce.

If we put sidespin on the ball we will not change the velocity of the ball in the direction that it is delivered. What we will do is to add a sideways velocity vector to the velocity in the x-axis. Although spin around both the x and the z-axis will cause side-spin when the ball pitches, spin around the z-axis causes sideways drift through the air, while spin around the x-axis will not. For spin around the z-axis the ball will drift in the opposite sense to the turn: off-spin will make the ball drift away from the batsman through the air and then turn towards the (right-handed) batsman on pitching, while leg-spin will drift into the batsman and then turn away from him on pitching.

When the ball is bowled with a combination of side and topspin, unusual effects can occur. In particular, when a legspinner bowls a googly the ball will almost always spin in both the x and the y-axis simultaneously; hence, the ball will turn into the batsman and, at the same time, hurry on and keep low, making it a big wicket-taking delivery for a bowler like Shane Warne. The different combinations of degree of spin in the two axes can cause multiple effects that are difficult for the batsman to anticipate, particularly if the pitch is responsive.

Why will a ball not turn on the first morning of a match, but will on the final afternoon? To get the ball to turn two things must happen. First, the bowler must get the ball to spin when he releases it from his hand; to do that he has to be able to grip the ball properly – if the ball is shiny and new it is difficult to grip it well enough to get a large number of revolutions on the ball. When the Indian side depended on its spinners to bowl the majority of the overs, Prasanna, Bedi, Chandrasekhar and Venkat relied on players like Eknath Solkar to bowl two or three overs at the start of the innings to rough up the ball and take some of the shine off it. Solkar, a middle order and sometimes opening batsman and brilliant close fielder, bowled in all but one of his 27 Tests, taking just 18 wickets at 59.4, but his purpose was simply to prepare the ball for the main strike force of his team: any wickets that he took were simply a bonus[11].

A typical Solkar effort was the Dehli Test v England in 1972. Opening the bowling with Abid Ali, he bowled just 3 overs in each innings, while Bedi and Chandrasekhar shared no less than 88.5 overs between them in the first innings alone. Once the new ball had been used a little and roughed-up, but was still hard, the spinners could get enough grip to spin the ball properly and also had the advantages of the harder ball with good bounce. As the ball gets older, and more roughed-up, it will stop slipping in the bowler's hand and achieve its maximum spin through the air. As its coefficient of friction increases it will also grip the pitch better and its softness will allow it to distort on pitching, increasing the area of contact with the ground. On the first day of a Test though, the pitch will normally have a reasonable covering of grass to bind it together, the pitch's coefficient of friction will be relatively low and the ball will have only brief contact with the surface over a small area of the ball, giving only a small degree of angular momentum transfer.

Over the five days of a Test the pitch receives a pounding from the ball. Any green that was present on Day 1 has usually gone completely on Day 5 and the bowlers have pounded the area of their follow-through with spikes. It is normal that the surface starts to loosen somewhat and that the footmarks become dusty and broken-up. If the ball lands in a footmark it will tend to bury itself a little in the loosened earth. A much larger surface of ball will make contact with the pitch, particularly as the ball gets softer and there will be greater friction giving a much larger angular momentum transfer from the ball. As a result the ball will tend to turn much more. A clever bowler though will vary constantly the amount of spin on the ball at delivery so that the batsman is never sure how much turn to play for.

A variant for the spin bowler is the arm ball. This is a ball that is not spun laterally on release, but is allowed to fly, seam up, with what backspin the natural release action gives. The idea is that, on pitching, the ball will follow the direction of the bowler's arm, without turning – hence the name. However, by being bowled seam up, some swing may be obtained that may trick the batsman into playing for non-existent turn in the opposite sense to the swing. After the bowler has delivered several balls that turn appreciably, the batsman instinctively expects another, similar ball and is inclined to play around it when the ball fails to turn. Of course, if the batsman anticipates the bowler's intentions he can play a shot with impunity, hence Vic Marks' complaint about being a "tortured genius" that opened this chapter.

One part of the game that has been lost, probably never to return, is a spinner bowling on drying pitch after heavy rain. This used to occur when pitches were left uncovered during the hours of play. Although a wet outfield is a major handicap, particularly to spin bowlers, as it makes the ball slippery and difficult to grip, a wet pitch will change into very viscous mud as it dries. This is the so-called *sticky dog*. The ball will adhere to the mud when pitching, allowing extreme angular momentum transfer and highly variable bounce. Good spin bowling can become almost unplayable in such conditions. A classic case was the Lords Test v Australia in 1934[12] when England scored 440 in their first innings. By the end of Day 2 (of four), Australia were in a strong position in reply at 192-2 and looking set to match or pass England's total. The following day was a rest day (Sunday) and the heavens opened. When Australia resumed batting on the Monday the pitch was starting to dry and, in five and a half hours, Australia lost 18 wickets for 210 runs, with Headley Verity taking 14 wickets in the day at a personal cost of 80 runs and finishing with match figures of 15-104[13]. Many people regret the passing of uncovered pitches and sticky dogs, pointing out that they were the supreme test for a batsman. The other side of the coin though is that often they turned batting into a lottery and matches into farces in which one side played under ridiculously inferior conditions and had no chance of avoiding defeat.

ENDNOTES

[1]   From "Three men in a car" in "Slices of Cricket" (Peter Roebuck, ISBN 0-04-796088-4)

[2]   Sri Lanka v England. 1ˢᵗ Test, 2003/04. Five other Sri Lankan bowlers managed combined figures of 69-13-169-3. He was then even more economical in the second innings, bowling 37 overs for 47 runs. Over the entire series Muralitharan bowled 231.4 overs and went for just 1.38 runs per over, taking a small matter of 26 wickets as he did so.

[3]   Having been dropped for the first two Tests of the tour of India, his reward was to be dropped again as Ashley Giles was fit again for England's next match, leaving him with a modest 8 wickets in his 4 Test career. Shaun Udal's First Class career also had an unusual end as he retired from playing for Hampshire towards the end of the 2007 season, only to be offered some months later the Middlesex captaincy for 2008 and come out of retirement to play three more seasons for Middlesex with considerable personal success.

[4]   This is the ultimate in embarrassing dismissals and has happened in the recent past to such players the former England opener Graeme Fowler and wicket-keeper Chris Read. However, it is a consequence of the innate difficulty for the brain in processing trajectory information when the ball is coming towards the batsman above eye level and is one of the reasons why bowling beamers is regarded as unfair.

[5]   Adams finally retired in 2008 after a final and unsuccessful appearance for Western Province in the SAA Provincial Three-Day Challenge Final (the South African Second Tier competition) in which he bowled just six overs, all in the second innings, in a losing cause. His final act before retirement was to collect a first ball duck as Griqualand West closed in on victory, with Western Province crashing from 162-5 to 174 all out chasing 217. He had even dropped out of First Class cricket completely for a time before returning for a last hurrah with Western Province, taking 13 wickets in five matches for them in his final season.

[6]   This was a truly bizarre match. Played over 3 days, as was the norm in England at the time, After Charles Bannerman fell for 4, McDonnell, Murdoch and Scott, the next three batsmen, scored 103, 211 and 102 respectively: a unique occurrence in Test cricket at the time. Lord Harris, the England captain, was nearing desperation as Australia reached 365-2 and ended up using all eleven players in his side to bowl – ten of them on the first day alone – also for the first time ever in a Test. Declarations were not allowed under the laws until 1889, so the Australian innings of 551 lasted no less than 311 four-ball overs. Lyttelton

handed the gauntlets to W.G. Grace when bowling, but kept his pads on to bowl and, at the end of each over, took the gloves back to continue as wicket-keeper.

England were 181-8 in reply on the third and final morning and in a desperate position when Lyttelton's wicket fell. Walter Read came in at 10 to join the opener, William Scotton, who was on 53 and proceeded to score a century in under two hours. The pair put on 151 together, with Scotton finally falling for 90 after almost six hours of self-denial and Read, last man out, for 117. Although England were forced to follow on, there was insufficient time left for Australia to obtain a result as only an hour of play remained.

No less than 535 overs were bowled in the three days. Although these were 4-ball overs, it is still equivalent to almost 120 overs per day. If this seems relatively little for a time when the pace of matches was much faster than today, remember that the hours of play were often shorter than now, with sometimes only 5 hours in a day's play.

You can read the Wisden report on the match at: http://www.espncricinfo.com/ wisdenalmanack/content/story/153432.html, the scorecard from the match at: http:// www.espncricinfo.com/ci/engine/match/62411.html and a short feature on Lyttelton's exploits at: http://www.espncricinfo.com/magazine/content/story/320591.html

[7] Conan Doyle, Arthur (1928), "The Story Of Spedegue's Dropper". Lightning Source Inc. ISBN 1425477208.

[8] Isaac Newton, "A letter of Mr. Isaac Newton, of the University of Cambridge, containing his new theory about light and colour," Philosophical Transactions of the Royal Society, vol. 7, pp. 3075-3087 (1671-1672).

[9] Greig's thirteen wickets in the 5th Test in Port of Spain allowed England to win the Test and draw a series in which the first three Tests had been completely one-way traffic.

[10] Bend it like Beckham.

[11] Solkar died tragically young at the age of 57 in 2005. Although described as "no more than an honest trier" as a bowler at Test match level, he achieved notoriety by dismissing Geoff Boycott cheaply four times in six innings during the 1974 Indian tour of England. Abid Ali, at least a more accomplished bowler, dismissed him on the other two occasions as Boycott aggregated just 58 runs in the three matches. The fallout from this led to Boycott's three year exile from Test cricket. His was a remarkable story. In his book "Marks out of

XI", Vic Marks recounts his story, Solkar being one of the few players to leave successfully the squalor endured by the Mumbai homeless.

12 http://www.espncricinfo.com/ci/engine/match/62620.html

13 A staggering (by today's standards) 361 overs were bowled in under 3 days of play. This match became legendary, not just for the bowling of Headley Verity. In particular it was the only Lords Test against Australia that England had won between 1896 and 2009. The Wisden match report is available at http://www.espncricinfo.com/wisdenalmanack/content/story/151792.html and some photographs, including the remarkable second innings dismissal of Don Bradman, at http://www.espncricinfo.com/ci/content/image/index.html?object=62620.

Headley Verity died as a prisoner of war after being mortally wounded and captured during the invasion of Sicily in the Second World War. Characteristically, his last words to his comrades were "keep going".

AFTERWORD

When I embarked on this project the Ashes 2010/2011 seemed a long way in the future. Nottingham University Press showed commendable faith in contracting the book and suggested a deadline for submission that seemed achievable, although I suggested a few extra months would be a good idea. Of course, 2010 turned out to be a very intense year, not just for cricket, but also for Herschel – the mission that I work for, with a profusion of late nights and long days, meaning that completion of the book took far longer than was intended, even though my employers have been supportive of my aspirations. It all means that events have, in some cases, overtaken the content of the book. I have resisted changing what I said in advance about the Ashes series with as much nobility as possible (no one, least of all me, predicted what was going to happen and how it would happen). Various controversial incidents in the Ashes series have shown that UDRS is not the universal panacea that some had hoped: it does not matter how sophisticated your technology is, some decisions will remain uncertain and a little common-sense is needed (is giving the batsman the benefit of the doubt in marginal cases really such a heinous crime?)

As this book is completed, India remain top of the ICC Test table and that position is getting steadily stronger, with South Africa, their nearest challenger, falling back after some disappointing results. It would be a brave man who would bet against India holding their position for at least the next two years. The divide though between the top five teams in Test cricket and the rest has become a gaping chasm. In the introduction to this book the background was the qualifying matches for the 2011 World Cup and, lo and behold, that World Cup is just around the corner, with as many as six teams entering it feeling that, with a little good fortune and a decent run of form, they could win the tournament. Looking at the longer term, the progress made by the Afghanistan team is so rapid that few would bet against the Afghans lighting up the 2015 World Cup with their enthusiasm (always assuming that the ICC relents on its plan to limit the World Cup only to the top teams and allows Associates to play - at the time of going to press this now seems extremely unlikely), despite all hardship and even possibly becoming a viable Test team within ten years. Cricket, that game that we love, goes on!

Over the course of writing the book I have had to research all sorts of details that I never imagined, from the properties of wood to the behaviour of viscoelastic substances, from botany to aerodynamics and from the history of Flemish weavers to the characteristics of ammunition. As the book progressed, it became obvious that "The Physics of Cricket" is an inadequate title, because cricket involves intimately many branches of science: not just physics, but engineering, chemistry and even botany. Some of the little gems of information uncovered have been included because they are fun facts (how many people know about that link between cricket balls and the American War of Independence???) Where information could not be obtained directly – no cricket ball manufacturer would even acknowledge my communications, let alone actually provide information – other, more indirect sources have been more forthcoming, such as manufacturers of cork composites and leather. Fans and experts in countries around the world have supplied information, opinions and critiques by e-mail or through on-line fora (it is amazing how diverse and heated opinions about umpiring and UDRS can be).

I hope that you enjoy this stroll through the Physics (and botany and mechanics and chemistry and ...) of cricket. The idea is to make it fun. Undoubtedly the critical reader will find things missing, or take issue with some of the content or conclusions, but will find things to interest him or her.

And, of course, the last word has to be to give my thanks to Cliff, Ros and especially to the patient and ever cheerful Sarah at NUP who has turned out this gorgeous manuscript from my Word files and to Trevor, across the pond in the USA, who got this whole project started.

INDEX

Player and Official Index

A

Abbas, Zaheer 150
Ackerman, Hylton 52
Adams, Paul 190, 191
Afridi, Shahid 58, 59, 64, 90, 190
Ahktar, Shoaib 85
Ahmed, Mushtaq 131
Akram, Wasim 131
Alam, Intikab 133, 187
Ali, Abid 198
Ali, Kabir 74
Amiss, Dennis 91
Anderson, James (Jimmy) 38, 161
Arafat, Yasir 67
Asif, Iqbal 146, 150
Atherton, Mike 149

B

Bannerman, Charles 9, 12, 21, 22, 201
Bari, Wasim 150
Bedi, Bishan 187, 198
Bedser, Alec 35, 139
Benaud, Richie 61, 132
Bird, Harold "Dickie" 167, 184
Bopara, Ravi 52, 162
Border, Allan 77, 78, 79, 188
Botham, Ian 61, 64, 77, 90, 111, 113, 120, 129
Boyce, Keith 14
Boycott, Geoff 34, 43, 75, 121, 126, 149, 182, 202
Boyle, Harry 11
Bradman, Donald (Don), Sir 34, 36, 37, 39, 41, 42, 43, 44, 49, 51, 53, 54, 57, 58, 113, 120, 127, 139, 140, 142, 203
Brearley, Mike 77, 90, 106, 111, 126
Bresnan, Tim 74
Broad, Stuart 107, 159

C

Caesar, Julius 7, 107
Cairns, Lance 66, 75
Cannings, Victor 56
Chandrasekhar, Bagwath 198
Chappell, Greg 111
Charles Dickens 8
Clarke, William 5
Close, Brian 61, 74
Cook, Alistair 59, 120, 121

D

Daniel, Wayne 14, 91
Dar, Aleem 185, 186
Davies, Winston 14
de Silva, Aravindra 114
Dev, Kapil 14, 65, 66
Dexter, Ted 105, 107
Dilley, Graham 64
Doctrove, Billy 163

E

Edmonds, Phil 107, 127
Emburey, John 77, 79, 80, 82, 84, 188
Erasmus, Marais 185

F

Fingleton, Jack 35
Finn, Steve 87

G

Ganguly, Sourav 184
Garner, Joel 79, 90